KU-489-874

Plant Carbohydrate
Biochemistry

Plant Carbohydrate Biochemistry

PROCEEDINGS OF THE
PHYTOCHEMICAL SOCIETY SYMPOSIUM
HERIOT-WATT UNIVERSITY
EDINBURGH, SCOTLAND
APRIL, 1973

Edited by

J. B. PRIDHAM

Department of Biochemistry
Royal Holloway College
University of London

1974

ACADEMIC PRESS
LONDON AND NEW YORK

ACADEMIC PRESS INC. (LONDON) LTD.
24/28 Oval Road,
London NW1

United States Edition published by
ACADEMIC PRESS INC.
111 Fifth Avenue
New York, New York 10003

Copyright © 1974 by
ACADEMIC PRESS INC. (LONDON) LTD.

All Rights Reserved
No part of this book may be reproduced in any form by photostat, microfilm, or any other
means, without written permission from the publishers

Library of Congress Catalog Card Number: 74-5655
ISBN: 0-12-564840-5

PRINTED IN GREAT BRITAIN BY
WILLIAM CLOWES & SONS LIMITED
LONDON, BECCLES AND COLCHESTER

Contributors

P. ALBERSHEIM, *Department of Chemistry, University of Colorado, Boulder, Colorado 80302, U.S.A.*

B. CHAPMAN, *Botany School, University of Cambridge, Cambridge, England.*

J. E. COURTOIS, *Department de Biochemie, Faculté des Sciences Pharmaceutiques et Biologiques, Paris, France.*

D. R. DAVIS, *Department of Biochemistry, Royal Holloway College, University of London, Englefield Green, Surrey, England.*

P. M. DEY, *Department of Biochemistry, Royal Holloway College, University of London, Englefield Green, Surrey, England.*

M. A. R. DE FEKETE, *Technische Hochschule Darmstadt, Fachbereich Biologie, Darmstadt, W. Germany.*

W. A. FULLER, *Botany School, University of Cambridge, Cambridge, England.*

A. HAUG, *Institute of Marine Biochemistry, University of Trondheim, 7034 Trondheim-NTH, Norway.*

H. KAUSS, *Fachbereich Biologie der Universität, 675 Kaiserslautern, Postfach 3049, W. Germany.*

B. LARSEN, *Institute of Marine Biochemistry, University of Trondheim, 7034 Trondheim-NTH, Norway.*

D. J. MANNERS, *Department of Brewing and Biological Science, Heriot-Watt University, Edinburgh, Scotland.*

B. W. NICHOLS, *Unilever Research Laboratories, Welwyn, Hertfordshire, England.*

D. H. NORTHCOTE, *Department of Biochemistry, University of Cambridge, Cambridge, England.*

J. B. PRIDHAM, *Department of Biochemistry, Royal Holloway College, University of London, Englefield Green, Surrey, England.*

T. AP REES, *Botany School, University of Cambridge, Cambridge, England.*

N. SHARON, *Department of Biophysics, The Weizmann Institute of Science, Rehovoth, Israel.*

B. STACEY, *Department of Chemistry, Sir John Cass School of Science and Technology, City of London Polytechnic, Jewry Street, London, England.*

R. J. STURGEON, *Department of Brewing and Biological Sciences, Heriot-Watt University, Edinburgh, Scotland.*

S. M. THOMAS, *Botany School, University of Cambridge, Cambridge, England.*

G. H. VIEWEG, *Technische Hochschule Darmstadt, Fachbereich Biologie, Darmstadt. W. Germany.*

C. L. VILLEMEZ, *Division of Biochemistry and Department of Chemistry, University of Wyoming, Laramie, Wyoming 82071, W. U.S.A.*

D. A. WALKER, *Department of Botany, The University, Sheffield, England.*

v

251560

Preface

This volume is a record of the Proceedings of the 1973 Annual General Meeting of the Phytochemical Society which was held at the Heriot-Watt University in Edinburgh, Scotland, during 4–6th April, 1973.

Edinburgh, a major centre of plant carbohydrate chemistry and biochemistry for many years, welcomed an international group of experts who lectured on topics ranging from CO_2 fixation to the biosynthesis of complex heteroglycans. Generous financial support for this meeting was given by the British Sugar Corporation, Ltd., Rank Hovis McDougall Research, Ltd. and Tate and Lyle Ltd.

The Phytochemical Society is indebted to Dr. J. S. D. Bacon, Professor J. E. Courtois, Professor Sir Edmund Hirst, F.R.S., Professor L. Hough and Dr. A. M. Macleod, who served as able chairmen of the various sessions, and to Professor D. J. Manners and Dr. R. J. Sturgeon, who were responsible for all the local arrangements.

The Editor must also thank all the contributors to the Symposium for their cooperation with regard to the publication of the Proceedings and Misses C. A. and W. J. Pridham who were most helpful with some items of the editorial work.

April 1974   J. B. PRIDHAM

Contents

CHAPTER 1

Introductory Address
Jean-Emile Courtois

CHAPTER 2

Some Characteristics of a Primary Carboxylating Mechanism
D. A. Walker

CHAPTER 3

**The Relationship between Gluconeogenesis and
Carbohydrate Oxidation in Higher Plants**
T. ap Rees, S. M. Thomas, W. A. Fuller and B. Chapman

CHAPTER 4

Plant Polyols
B. E. Stacey

CHAPTER 5

Some Aspects of Sucrose Metabolism
D. R. Davies

CHAPTER 6

The Nature and Function of Higher Plant α-Galactosidases
J. B. Pridham and P. M. Dey

CHAPTER 7

The Structure and Function of Plant Glycolipids
B. W. Nichols

CHAPTER 8

Some Aspects of the Enzymic Degradation of Starch

D. J. Manners

CHAPTER 9

Starch Metabolism: Synthesis Versus Degradation Pathways

M. A. R. de Fekete and G. H. Vieweg

CHAPTER 10

**The Primary Cell Wall and Central Control of
Elongation Growth**

Peter Albersheim

CHAPTER 11

**Sites of Synthesis of the Polysaccharides of the
Cell Wall**

D. H. Northcote

CHAPTER 12

The Relation of Plant Enzyme-Catalysed β-(1,4)-Glucan Synthesis to Cellulose Biosynthesis *in vivo*

C. L. Villemez

CHAPTER 13

Biosynthesis of Pectin and Hemicelluloses

Heinrich Kauss

CHAPTER 14

Biosynthesis of Algal Polysaccharides

Arne Haug and Bjørn Larsen

CHAPTER 15

Chemical and Biochemical Aspects of Fungal Cell Walls

R. J. Sturgeon

CHAPTER 16

Glycoproteins of Higher Plants
Nathan Sharon

CHAPTER 1

Introductory Address

JEAN-EMILE COURTOIS

Département de Biochimie, Faculté des Sciences
Pharmaceutiques et Biologiques, Paris, France

It is a very great honour for me to give this address for two reasons. The first is that I have been working since 1930 on carbohydrates, and more especially those of Phanerogams. The second reason is that the Symposium is being held in Scotland and in particular in this town of Edinburgh, where many carbohydrate chemists of world-wide repute have worked.

We must always keep in mind that carbohydrates have played a leading part in the advancement of civilization. An essential step in the evolution of mankind was the beginning of agriculture, which occurred as an integral part of the change from the paleolithic to the neolithic period. At this time man selected for cultivation plants rich in starch, such as wheat and barley in the Middle-East (which spread to Europe and North Africa), rice in Eastern Asia and maize in Central America.

Until the end of the last century, the possession of areas suitable for growing such crops was the reason for many invasions and wars. In addition, until the nineteenth century, wood, with its high percentage of carbohydrates, remained the raw material for framework of houses, furniture, and as I am amongst seafaring people, I must also mention boats.

Plant carbohydrates are still of great economic importance as sources of food, paper and gums. Hence it follows that research on carbohydrates has more or less immediate consequences on our way of life. This research is now located at the convergence point of modern methods in organic chemistry, enzymology, biophysics and physiology.

In the study of a plant carbohydrate, the main line of approach is usually as follows: isolation, purification, determination of constitutive monosaccharides and examination of their modes of linkage. This can be extended to several other types of research including plant taxonomy, polymer conformations and enzyme-catalysed reactions. The latter leads to an understanding of the biosynthesis and catabolism of the compound.

As an example of this evolution, I shall discuss the galactomannans from leguminous seeds. They accumulate in the cells of the endosperm as the seeds mature. In the past these polysaccharides were considered to be hemicelluloses, gums or mucilages: at the present time, however, they are normally classified as a separate, specific group of polymers.

Some of the galactomannans have been used for more than 5000 years on account of their characteristic viscosities and their ability to form films on desiccation. The coverings on the Egyptian mummies were impregnated with guar and locust bean gums: for this reason Rees (1972) humorously suggested that they be named "Pharaoh's polysaccharides".

Isolation and Determination of Constituent Sugars

Bourquelot and Hérissey (1899) showed that gums from leguminous seeds on acidic hydrolysis yielded galactose and, at that time a new aldohexose, mannose. They noticed that during germination the seeds produced a *seminase* (a mixture of glycosidases), which eventually split the polysaccharides into their two constitutive aldohexoses.

Bourquelot and Hérissey (1900) also observed that the molar ratio of mannose to galactose was constant in a particular species of seed, but that there were differences between species. This ratio has now been determined for seeds from more than 40 plants; it ranges from 1·0 for alfalfa and clover to 3·8 for locust bean and *Gleditschia* and is as high as 5·2 for *Sophora*. There is a relationship between this ratio and the classification of legume sub-families (Reid and Meier, 1970b). This is a new demonstration of the growing importance of the use of chemotaxonomy in botany.

Glycosidic Linkages

One of the first studies of a galactomannan structure was made by Whistler and his group (Whistler and Smart, 1953) on guar gum. Partial acid hydrolysis provided mannose oligosaccharides (DP 2-5) joined by β-(1 → 4)-linkages, and 6-*O*-D-galactosyl-D-mannose. Permethylation followed by hydrolysis and periodate oxidation studies have confirmed these data.

The classical methods for the establishment of polysaccharide structure have since been applied to about 25 galactomannans.

Galactomannans from leguminous seeds form a rather homogeneous group with the same basic structure, i.e. a β-(1 → 4)-linked D-mannopyranose backbone with the C-6 positions of most of the mannose residues substituted by single α-D-galactopyranosyl units. In a few cases the majority of the mannose residues are substituted in this way.

Unusual exceptions to this structure are known. For example, some β-(1 → 3)-linked D-mannose residues have been found in the galactomannans from *Leucaena glauca* (Unrau, 1961) and *Caesalpinia pulcherrima* (Unrau and Choy, 1970) and in the *Gleditschia* polysaccharide, a small number of the side chains consist of α-(1 → 6)-linked D-galactose disaccharide units bonded to C-6 positions on the mannose backbone (Courtois and Le Dizet, 1963; Cerezo, 1965; Leschziner and Cerezo, 1970).

When the molar ratio of mannose to galactose in galactomannans is higher than 1·5, β-mannanases split some parts of the chain (Courtois and Le Dizet, 1968, 1970). The enzyme hydrolyses the glycosidic bonds of mannose residues which are not substituted with galactose. The product is a core in which all the mannosyl units are linked through position 6 to galactosyl residues. For these galactomannans, there is a chain of non-branched zones alternating with zones where all the mannosyl units are branched according to the following structure:

$$\text{Gal } p \ \alpha\text{-1}$$
$$\downarrow$$
$$6$$
$$[\text{Man } p \ \beta\text{-}(1 \rightarrow 4)]^m - [\text{Man } p \ \beta\text{-}(1 \rightarrow 4)]^n$$

The ratio m/n is consistent with the molar ratio mol mannose/mol galactose.

Conformation Studies

Galactomannans are water dispersible and produce highly viscous solutions, consequently they are important industrial materials, particularly in the food industry. Many factors affect viscosity: an increase in temperature, for example, disrupts loose aggregates which do not re-associate completely on cooling.

Interaction of galactomannans possessing a low galactose content with solutions of agarose or κ-carrageenan has systematically been studied by Rees (1972) and Dea *et al.* (1972). The non-substituted mannosyl units of the galactomannan are believed to link to the helical part of the polysaccharides and these combinations facilitate gel formation. It appears that hydrogen bonding between the two polysaccharides occurs and there is a change of quaternary structure. Dea *et al.* (1972) suggest that the binding might mimic biological cohesion between skeletal and gel phases of natural cell walls.

Hydrolysis by Glycosidases

α-Galactosidases easily hydrolyse phenolic α-D-galactosides; *p*-nitrophenyl α-D-galactoside, for example, is commonly used as a substrate for these enzymes (e.g. Dey and Pridham, 1972). The specificity of α-galactosidases from different plant species is variable; some of them, for example, hydrolyse raffinose and galactomannans whereas others are specific for low molecular weight substrates (Courtois *et al.*, 1958). In our laboratory, we have purified several α-galactosidases from coffee bean and from germinated alfalfa and fenugreek seeds which can remove galactose from galactomannans. When coffee bean α-galactosidase is allowed to react with *Gleditschia* galactomannan, which has a molar ratio of mannose to galactose of 3·8, an insoluble product results with a ratio which has increased to 28 (Courtois and Le Dizet, 1966). The structure of the degraded product is similar to that of mannan from fruits of palm trees and some of its β-D-mannosidic linkages can be hydrolysed by an endo-β-(1 → 4)-mannanase from germinated fenugreek seeds (Clermont-Beaugiraud and Percheron, 1968).

Biogenesis

In *Gleditschia*, the mannose/galactose ratio in galactomannans is approximately the same in both green and mature seeds (Courtois and Le Dizet, 1963). The same is true for fenugreek seeds (Reid and Meier, 1970a). The galactomannan begins to be formed at an early stage of seed development, and then increases in amount throughout the growth of the seed with no change in chemical composition. UDP-galactose is probably the source of galactose in galactomannans and the α-D-galactoside of myoinositol (galactinol) is possibly an intermediate in the galactosylation reaction.

Function

Galactomannans seem to have a double physiological function in seeds: (1) they retain the water by solvation and hence prevent complete drying out which could induce protein denaturation, especially of enzymes necessary at the onset of seed germination. The galactose side chains constitute the hydrophilic part of the galactomannan whereas the mannose backbone is hydrophobic, as it is in the case of mannans from palm-nuts; (2) they are food reserves by virtue of the fact that they can readily be hydrolysed to their two constitutive hexoses.

Utilization

Reid (1971) and Reid and Meier (1972), who have studied germination of fenugreek seeds using electron miscroscopy and chemical analysis, have shown that the aleurone layer is the centre of intensive protein synthesis and produces hydrolytic enzymes acting upon galactomannans. The resulting hydrolysis products can then be used for synthesizing starch in the cotyledons.

During germination there is a marked rise in α-galactosidase activity and the galactose released is presumably converted to D-galactose-1-phosphate by galactokinase (Sioufi *et al.*, 1970).

β-Mannanase activity, however, is low in germinated seeds. Studies by Foglietti and Percheron (1972) suggest that hydrolysis by β-mannanase is not the only metabolic process occurring during galactomannan utilization by the germinating seed. These workers have succeeded in preparing a new phosphorolytic enzyme from germinated seeds of fenugreek. The enzyme (β-(1 → 4)-mannan:orthophosphate β-D-mannosyltransferase) catalyses the following reversible reaction which allows either the transfer of mannose from β-D-mannosyl-1-phosphate to the growing chain or phosphorolysis of the polymer:

$$(\text{mannose})_n + \beta\text{-D-mannose-1-phosphate} \rightleftharpoons (\text{mannose})_{n+1} + \text{orthophosphate}$$

The two hexose phosphates produced from galactomannans can be converted to the corresponding nucleotide sugars (Sioufi *et al.*, 1970). During the germination of fenugreek seeds, for example, there is a noticeable rise in the level of UDP-glucose, UDP-galactose and GDP-mannose. The nucleotide sugars, in turn, can be reconverted to hexose-1-phosphates and nucleoside diphosphates by a nucleotide pyrophosphatase which was recently identified by Clermont *et al.* (1973). The enzyme is more reactive with GDP-mannose and UDP-galactose than UDP-glucose. An epimerase, catalysing the conversion of UDP-galactose to UDP-glucose has also recently been identified in fenugreek seeds by F. Percheron and his collaborators (unpublished).

Conclusions

Using the galactomannans as an example, we can follow the evolution of research, starting with the working out of the structure of a polysaccharide, which opened the way to a wide range of other studies dependent on biophysics, enzymology and plant physiology at the cellular level.

The papers that follow will, no doubt, confirm that carbohydrates are still in the limelight of scientific advancement and this is very satisfying for chemists who study their structures.

REFERENCES

Bourquelot, E. and Hérissey, H. (1899). *J. Pharm. Chim.* **10**, 438.
Bourquelot, E. and Hérissey, H. (1900). *J. Pharm. Chim.* **11**, 589.
Cerezo, A. A. (1965). *J. Org. Chem.* **30**, 924.
Clermont-Beaugiraud, S. and Percheron, F. (1968). *Bull. Soc. Chim. Biol.* **50**, 41.
Clermont, S., Foglietti, M. J. and Percheron, F. (1973). *C. R. Acad. Sc. Paris* **276**, 843.
Courtois, J. E., Anagnostopoulos, C. and Petek, F. (1958). *Bull. Soc. Chim. Biol.* **40**, 1277.
Courtois, J. E. and Le Dizet, P. (1963). *Bull. Soc. Chim. Biol.* **45**, 731.
Courtois, J. E. and Le Dizet, P. (1966). *Carbohyd. Res.* **3**, 141.
Courtois, J. E. and Le Dizet, P. (1968). *Bull. Soc. Chim. Biol.* **50**, 1695.
Courtois, J. E. and Le Dizet, P. (1970). *Bull. Soc. Chim. Biol.* **52**, 15.
Dea, I. C. M., McKinnon, A. A. and Rees, D. A. (1972). *J. Mol. Biol.* **68**, 153.
Dey, P. M. and Pridham, J. B. (1972). *Adv. Enzymol.* **36**, 91.
Foglietti, M. J. and Percheron, F. (1972). *C. R. Acad. Sc. Paris* **274**, 130.
Leschziner, C. and Cerezo, A. S. (1970). *Carbohyd. Res.* **15**, 291.
Rees, D. A. (1972). *Biochem. J.* **126**, 257.
Reid, J. S. G. and Meier, H. (1970a). *Phytochemistry* **9**, 513.
Reid, J. S. G. and Meier, H. (1970b). *Z. Pflanzenphysiol.* **62**, 89.

Reid, J. S. G. (1971). *Planta* **100**, 131.

Reid, J. S. G. and Meier, H. (1972). *Planta* **106**, 44.

Sioufi, A., Percheron, F. and Courtois, J. E. (1970). *Phytochemistry* **9**, 991.

Unrau, A. M. (1961). *J. Org. Chem.* **26**, 3097.

Unrau, A. M. and Choy, Y. M. (1970). *Carbohyd. Res.* **14**, 151.

Whistler, R. L. and Smart, C. L. (1953). "Polysaccharide Chemistry", pp. 296–299. Academic Press, New York and London.

CHAPTER 2

Some Characteristics of a Primary Carboxylating Mechanism

D. A. WALKER

Department of Botany, The University, Sheffield, England

I. INTRODUCTION

Almost 20 years have elapsed since the elucidation (see e.g. Bassham and Calvin, 1957) of what is now often referred to as the Benson-Calvin cycle and it might seem surprising that anyone should seek, at this late stage, to reiterate the essential characteristics of a primary photosynthetic carboxylation mechanism. However, recent events have indicated that the path of carbon is not necessarily the same in all higher plants (see e.g. Hatch, 1970; Hatch and Slack, 1970) and for this reason it was felt that the present exercise might be usefully undertaken. The intention is to list certain characteristics which, in theory, would be entirely necessary or even highly desirable in a primary carboxylating mechanism and to examine the extent to which these are actually realized in the Benson-Calvin cycle (i.e. in the C-3 pathway of photosynthesis) and whether or not these characteristics are also features of "C-4 photosynthesis" and Crassulacean Acid Metabolism (CAM).

II. A Primary Carboxylating Mechanism Defined

"Carboxylation" is a process in which CO_2 is joined to an existing molecule in such a way that a new carboxyl group is formed. "Primary" is used in the dictionary sense of, "earliest, original, of the first rank of a series" and especially, "not derived". In the latter sense a primary carboxylating mechanism must be a self-sufficient breeder reaction from which all other secondary carboxylations, and indeed all other metabolic events, are ultimately derived. Such a mechanism should incorporate most or all of the features listed in Section III.

III. Essential and Desirable Characteristics of a Primary Carboxylating Mechanism

A. THE EQUILIBRIA OF THE CARBOXYLATIONS

Higher plants are normally exposed to atmospheres containing approximately 0·03% (300 ppm) CO_2 or to solutions in equilibrium with such atmospheres. From such atmospheres carbon dioxide has to move along a concentration gradient to the site of carboxylation and there, in the absence of some concentrating device or CO_2 pump, its concentration will be lower than it is outside, possibly as low as 50–100 ppm (the intercellular CO_2 concentration of many photosynthesizing C-3 plants lies within this range). As a biological process, photosynthesis is rapid and the accumulation of photosynthetic products is considerable. End product inhibition could be offset by equally rapid and substantial removal of products from the site of formation but it is evident that metabolic advantages would result if the carboxylation itself possessed a favourable equilibrium position. A carboxylation, as such, is an energy-consuming process since it involves the conversion of a highly stable carbon dioxide molecule to a somewhat less stable carboxyl group. For example the decarboxylation of oxaloacetate to pyruvate (reaction 1)

$$\text{Oxaloacetate} \rightarrow \text{pyruvate} + CO_2 \tag{1}$$

is accompanied by a decrease in free energy of approximately -6 kcal (Walker, 1962) and as a first approximation the energy required to "drive" a carboxylation may be taken as approximately equal to that released by the hydrolysis of ATP. Phosphoenolpyruvate is a high energy substrate in this sense. Its hydrolysis to pyruvate and orthophosphate yields about 14 kcal so that its hydrolytic carboxylation (reaction 2) catalysed by phosphoenolpyruvate carboxylase yields approximately 7 kcal (Walker, 1962).

$$\text{Phosphoenolpyruvate} + CO_2 + H_2O \rightarrow \text{oxaloacetate} + \text{orthophosphate} \tag{2}$$
$$(\Delta F^b \simeq -7 \text{ kcal})*$$

* ΔF^b is the change in free energy calculated for 25°C, pH 7·0, 0·05% CO_2 and other reactants at 0·01 M.

This gives $K > 10^5$ so that at equilibrium the ratio of substrates to products is in excess of $1:100\ 000$ and if carboxylation is followed by reduction to malate (with an equally favourable equilibrium position) the ratio of products to substrate (P/S) increases to approximately 10^{11}.

The reaction catalysed by ribulose diphosphate carboxylase (reaction 3) which is also a hydrolytic carboxylation, is marginally more favourable than that catalysed by phosphoenolpyruvate carboxylase (Bassham and Krause, 1969).

$$\text{Ribulose-1,5-diphosphate} + CO_2 + H_2O \rightarrow 2(\text{3-phosphoglycerate}) \quad (3)$$
$$(\Delta F^b \simeq -8 \text{ kcal})$$

Thermodynamically, both reactions are eminently suitable for fixing CO_2 and would abstract CO_2 from normal atmospheres in the presence of an appropriate catalyst.

At one time the above remarks would have been considered unexceptionable and indeed thermodynamics are still generally held to be of importance in assessing what could be called the "energetic feasibility" of a reaction or reaction sequence. Nevertheless, it is true that (in the strictest sense) classical thermodynamics can only be applied to closed systems and that intermediary metabolism is an open system in which reactants are frequently replenished and products are frequently removed. This has led some workers to question the relevance of free energy changes in this sort of context and, for example, Banks and Vernon (1970) state that the behaviour of open systems "is not derivative of standard free energy (i.e. of equilibrium constants)" and that "synthesis can occur irrespective of whether the equilibrium constant is highly unfavourable provided only that the products are effectively removed". These workers recognize that there is some limitation but say, "provided that the standard free energy change associated with a reaction, which is part of a sequence, is not too largely positive the reaction will occur given efficient catalysis". While accepting this point in principle it seems necessary to seek some qualification of "not too largely positive" and "efficient catalysis".

The reaction catalysed by malic dehydrogenase illustrates this point.

$$\text{oxaloacetate} + NADH + H^+ \rightarrow \text{malate} + NAD^+ \quad (4)$$

The equilibrium position ($\Delta F \simeq -7$ kcal, as written) overwhelmingly favours oxaloacetate reduction (Krebs, 1954) and in the closed system constituted by the spectrophotometer cell there is no appreciable oxidation of malate at other than abnormal substrate concentration and pH. Conversely in the open system of the Krebs cycle the reaction proceeds freely as the end products are removed and the reactants replenished (Krebs, 1954). In this instance it is clear that the catalysis is "efficient" and that the equilibrium can be displaced in practice as well as in theory. In other circumstances this is apparently not so. When pyruvate and $^{14}CO_2$ were incubated with oxaloacetic decarboxylase and malic dehydrogenase no radioactivity was recovered in malate (Herbert, 1951) even though the overall equilibrium position ($\Delta F^b \simeq -2$ kcal) of the combined reactions (5 and 6) is not unfavourable.

$$\text{pyruvate} + CO_2 \rightarrow \text{oxaloacetate} \qquad (\Delta F^b \simeq 6 \text{ kcal}) \tag{5}$$

$$\text{oxaloacetate} + NADH + H^+ \rightarrow \text{malate} + NAD \qquad (\Delta F^- \simeq -8 \text{ kcal}) \tag{6}$$

Moreover, the overall reductive carboxylation of pyruvate to malate (reaction 7) by malic enzyme (Ochoa *et al.*, 1947) is well known.

$$\text{pyruvate} + CO_2 + NADPH + H^+ \rightarrow \text{malate} + NADP \tag{7}$$

Presumably the lack of detectable incorporation of radioactivity into malate in the presence of oxaloacetic decarboxylase and malic dehydrogenase can be attributed to inefficient catalysis. That is, the rate of formation of oxalo-acetate brought about by the first enzyme was insufficient, under the conditions employed, to permit the second enzyme to operate an effective sink and hence displace the unfavourable equilibrium position of the first stage (cf. Walker, 1962). It is true that the success or otherwise of this reductive carboxylation is governed by kinetic parameters but the kinetics are evidently not themselves independent of the thermodynamics. For example, if reaction 5 is replaced by that catalysed by pyruvic carboxylase (reaction 8), so that the free energy change is shifted from $+6$ to -1, then the formation of malate occurs readily in the overall sequence in which reaction 8 is followed by reaction 6.

$$\text{pyruvate} + CO_2 + ATP \rightarrow \text{oxaloacetate} + ADP + \text{orthophosphate} \atop (\Delta F^b \simeq -1 \text{ kcal}) \tag{8}$$

Similarly, reaction 2 combined with reaction 6 is even more favourable. In short, although it might be theoretically possible to have a primary carboxylation with an unfavourable equilibrium position, in practice a favourable equilibrium position will clearly lessen the problems associated with the rapid synthesis of concentrated product from dilute substrate (see also Walker, 1962). In this respect there is nothing to choose between the two carboxylations (reactions 2 and 3) which are common to C-3 and C-4 photosynthesis as well as to Crassulacean Acid Metabolism.

B. THE AFFINITY OF THE CARBOXYLASES FOR CO_2/BICARBONATE

1. Concentration of CO_2 for Half Maximal Velocity

Although a favourable equilibrium position is important it would be relatively ineffective if the carboxylase had a low affinity for CO_2. In this respect phosphoenolpyruvate carboxylase has been acceptable since its high affinity was first measured in 1956 (Walker, 1957; Walker and Brown, 1957) but the apparently low affinity of ribulose diphosphate carboxylase (Weissbach *et al.*, 1956) constituted a major problem until very recently. Thus although the evidence giving this enzyme a central role in photosynthesis was compelling in almost every other respect, the fact that it seemed to need something in the region of 1·8–22% CO_2 in the gas phase to reach half maximal velocity virtually precluded the possibility that it could act unaltered or unaided (see Walker, 1973). The demonstration that the reaction involved CO_2 rather than bicarbonate (Cooper et al., 1969) eliminated the higher of the two values given above but did nothing in itself to resolve the real problem. (At pH 7·9 and 25°C, 1·8% atmospheric CO_2 is in equilibrium with 22 mM bicarbonate and 0·54 mM CO_2 in solution: see Walker, 1973.) What did help was the demonstration in several laboratories that Mg^{2+} ion depresses the K_m (CO_2) and increases the V_{max} (Sugiyama et al., 1968, 1969; Bassham et al., 1968; Lin and Nobel, 1971; Jensen, 1971; Stokes et al., 1971; Walker et al., 1971; Walker, 1971). For the isolated enzymes this increase in affinity is only 4- to 15-fold but if the enzyme is still in situ within a carefully ruptured chloroplast (Jensen, 1971) the increase in affinity may be as much as 50-fold so that half maximal velocity is realized in solutions which would be more or less in equilibrium with the levels of CO_2 in the atmosphere. Why the K_m is lower in situ has still to be established but two possibilities are considered in Sections B2 and B3 below.

2. Light Activation

Light activation and dark deactivation of the carboxylation is desirable to the extent that it would conserve substrate (see Walker, 1973). In the absence of such a switch an equilibrium constant of 10^6 would ensure virtually complete utilization of the substrate for the carboxylase in the dark and this, in turn, would necessitate rapid, light initiated mobilization of some reserve in the next light period. Whether or not such control is an absolute necessity or a useful refinement is difficult to say. However, it is evident that the regeneration of the CO_2-acceptor is directly dependent on ATP and NADPH (and hence subject to light intensity) and if carboxylation were entirely independent of light, disadvantages could ensue as a consequence of consumption and regeneration becoming unbalanced. Certainly CO_2 fixation by ribulose diphosphate carboxylase stops in the dark long before the substrate is exhausted (Bassham and Jensen, 1966; Jensen and Bassham, 1968) and this sort of observation, together with measurements of magnesium fluxes into the stromal compartment has led to the proposal that the light/dark switch is activated by Mg^{2+} (Fig. 1) through its known effect on the K_m and V_{max} (see review by Walker, 1973). Simulation of Mg^{2+} activation has been demonstrated in a reconstituted chloroplast system (Walker et al., 1971; Walker, 1973) in which CO_2-dependent O_2 evolution can be "switched on" by increasing the $MgCl_2$ concentration from 1 to 5 mM and from 1 to 3 mM (increases from near zero to 11 mM have been held to be feasible on the basis of observed Mg^{2+} fluxes and observed volume changes in intact illuminated chloroplasts: Lin and Nobel, 1971).

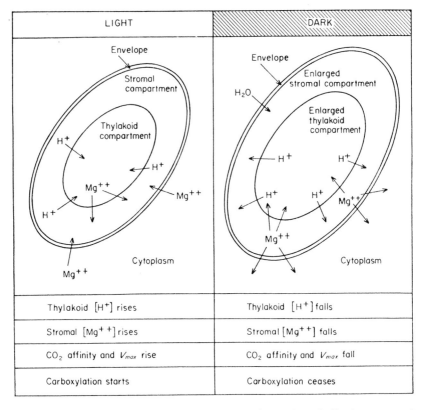

Fig. 1. Light activation of ribulose diphosphate carboxylation. Stylized representation of the events which are believed to bring about changes in ribulose diphosphate carboxylase in the light such that its affinity for CO_2 increases by a factor of 50 or more. Photosynthetic electron transport brings about a charge separation and an influx of protons into the thylakoid compartment. Associated with these changes there is an influx of Mg^{2+} ions which enter the stroma from the thylakoids and also from the cytoplasm. The increased Mg^{2+} level alters the affinity of the enzyme for CO_2 and also the V_{max} and shifts its pH optimum in an acid direction. These processes are reversed in the dark. (From Walker, by courtesy of MTP Reviews, in press.)

Examples of such Mg^{2+} activation are given in Figs 2 and 3. In Fig. 2 the only initial difference between the two reaction mixtures is that one (A) contains 4 mM $MgCl_2$ and the other (B) only 1 mM $MgCl_2$. It will be seen that, in the presence of the higher Mg^{2+} concentration, oxygen evolution was more or less continuous but in the lower concentration it ceased as soon as the added NADP (0·2 μmol) had been reduced. It could then be re-started, following a lag, by the addition of $MgCl_2$ to bring the concentration in B to 4 mM. In Fig. 3 the conditions were much the same as in Fig. 2 except for the presence of a little radioactive bicarbonate and the course of O_2 evolution was, therefore, also very similar, showing Mg^{2+} activation. In this experiment, however, CO_2-fixation was measured simultaneously and can be seen

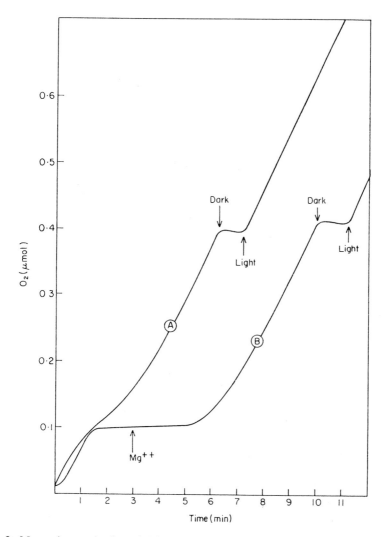

FIG. 2. Magnesium activation of CO_2-dependent oxygen evolution. Both 2 ml reaction mixtures contained ruptured chloroplasts (100 μg chlorophyll) stromal enzymes (approx. 5 mg protein), NADP (0·1 μmol), ATP (4 μmol) ribose-5-phosphate (2 μmol) sodium bicarbonate (10 μmol) ferredoxin (approx. 1·5 μmol) glucitol (0·33 M), N-2-Hydroxyethyl-piperazine-N'-2-ethanesulphonic acid (50 mM at pH 7·6) and EDTA (2·0 mM). In addition $MgCl_2$ (8 μmol) was added to reaction mixture A from the outset and to reaction mixture B as indicated. Illumination was started a few seconds after zero time and the initial oxygen evolution is associated with the reduction of NADP and, therefore, ceases in B because the CO_2-dependent generation of 3-phosphoglycerate (which reoxidizes NADPH as it is itself reduced to triose phosphate) is "switched-off" (see text and Fig. 1). Following the addition of $MgCl_2$ to B, carboxylation starts and is followed by oxygen evolution as 3-phosphoglycerate accumulates.

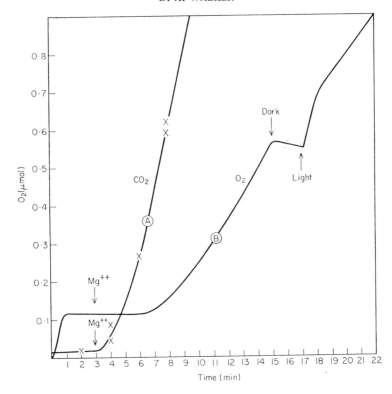

FIG. 3. Magnesium activation of CO_2 fixation and CO_2-dependent oxygen evolution. Oxygen evolution and CO_2 fixation were measured simultaneously in a reaction mixture which was essentially the same as those in Fig. 2 except for the additional presence of $NaH^{14}CO_3$ (200 μCi). It will be seen that the course of O_2 evolution (curve B) was very similar to that in Fig. 2b except for the consequences of including a dark interval. This interval stopped O_2 evolution but when illumination was continued evolution was at first renewed at an increased rate because, in this instance, of the accumulation of 1,3-diphosphoglycerate in the dark (the mixture contains substrate quantities of ATP so that CO_2 fixation is not dependent on photosynthetic electron transport). CO_2 fixation (curve A) was negligible until the addition of $MgCl_2$ but then commenced very rapidly, indicating that the much longer lag in O_2 evolution must be largely associated with the factors concerned in the formation of an optimal steady state concentration of 1,3-diphosphoglycerate.

to respond almost immediately to the addition of $MgCl_2$ indicating that much of the lag in the oxygen response may be attributed to the time which must elapse before 3-phosphoglycerate (derived from CO_2 fixation) reaches the steady state concentration necessary to maintain maximal reoxidation of reduced NADP (cf. Walker, 1973).

Phosphoenolpyruvate carboxylase (which, incidentally, is believed to utilize bicarbonate rather than CO_2 (Cooper and Wood, 1971) does not undergo rapid light activation but phosphate pyruvate dikinase (which is involved in regeneration of pyruvate in C-4 plants) does.

3. *Bicarbonate–CO_2 Pump*

In Section B2 the possibility was considered that light activation of ribulose diphosphate carboxylase was mediated by movements of Mg^{2+} into the stroma. If this process actually occurs it seems likely that the Mg^{2+} movements would be linked, directly or indirectly, with the establishment of the light-driven pH gradient. This gradient would also tend to make the stroma more alkaline and thus favour the formation of bicarbonate ion from CO_2. This would constitute a bicarbonate pump (for which there is direct evidence: Werdan and Heldt, 1973) but it would not increase the stromal CO_2 concentration (CO_2 solubility is largely independent of pH but the bicarbonate concentration in equilibrium with dissolved CO_2 increases considerably as the pH is raised from 5 to 9). An ingenious proposal by Werdan and Heldt (1973) would nevertheless allow this increase in bicarbonate to accelerate the rate of ribulose diphosphate carboxylation (see Fig. 4). This reaction utilizes CO_2 rather than bicarbonate and Werdan and Heldt propose that in the micro-environment at the carboxylase surface the

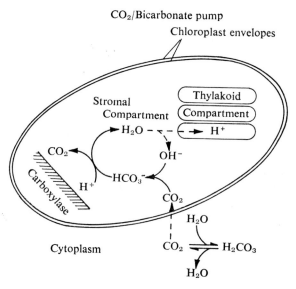

FIG. 4. Proposed CO_2-bicarbonate pump. Photosynthetic electron transport initiates a pH gradient as protons are moved from the stromal compartment into the thylakoids. The stroma becomes more alkaline and bicarbonate formation (catalysed by carbonic anhydrase) is favoured. CO_2 passes inwards through the chloroplast envelopes to maintain the CO_2–bicarbonate equilibrium. At the ribulose diphosphate carboxylase surface carboxylation causes local acidification as two molecules of 3-phosphoglycerate are formed from one molecule of ribulose-1,5-diphosphate and one molecule of CO_2. This local acidification favours the release of CO_2 from bicarbonate (also catalysed by carbonic anhydrase). It may be noted that Mg^{2+} activation (Figs 1–3) shifts the pH optimum towards the acid side and that this would clearly be consistent with the above proposal. This figure is largely based on a proposal by Werdan and Heldt (1973).

formation of acid (3-phosphoglycerate) would give rise (by acidification of bicarbonate in a carbonic anhydrase-catalysed reaction) to a local increase in free CO_2 (cf. Poincelot, 1972). This notion is supported by the fact that a ten-fold increase in Mg^{2+} can shift the pH optimum of the carboxylase about 1 unit to the acid side (whereas the pH of the stroma as a whole would be expected to become more alkaline under conditions which would favour an increase in the Mg^{2+} concentration). Both of these facts can also be related to the K_m (CO_2) for ribulose diphosphate carboxylase which is lower, by a factor of 10, when assayed *in situ*. Interaction of carboxylase sub-units might be favoured in the undisturbed protein and this could affect the K_m but alternatively the low K_m *in situ* could indicate the operation of the processes suggested by Werdan and Heldt (1973). Thus, in order to achieve a given CO_2 concentration it is necessary to add approximately 9 times less bicarbonate at pH 6·5 than at 7·5. Clearly, if the enzyme within the chloroplast were working at pH 6·5 when the external medium was at 7·5 the lower K_m recorded by Jensen (1971), which increases 10-fold upon agitation, could be explained.

C. REGENERATION OF THE CO_2 ACCEPTOR

All known photosynthetic carbon fixation occurs by carboxylation, i.e. CO_2 is joined to an existing acceptor forming a new carboxyl group. If this process is to continue the CO_2 acceptor must be regenerated. In the Benson-Calvin cycle this is achieved (Fig. 5) by a rearrangement of 5 molecules of triose phosphate to yield 3 molecules of pentose monophosphate which are, in turn, converted and phosphorylated to yield ribulose diphosphate.

In C-4 photosynthesis (Hatch, 1970) phosphoenolpyruvate is regenerated from pyruvate in reaction 9, which is catalysed by phosphate pyruvate dikinase (see Fig. 8).

$$\text{pyruvate} + \text{ATP} + P_i \rightarrow \text{PEP} + \text{AMP} + PP_i \qquad (9)$$

This reaction is not itself energetically more favourable than that catalysed by pyruvate kinase (reaction 10),

$$\text{pyruvate} + \text{ATP} \rightarrow \text{PEP} + \text{ADP} \qquad (10)$$

but when linked to pyrophosphate hydrolysis (reaction 11),

$$PP_i + H_2O \rightarrow P_i + P_i \qquad (11)$$

the net effect is to utilize both high energy groups of ATP so that the overall sequence (reaction 10 plus 11) has a free energy change of approximately zero. The pyrophosphatase has a sufficiently low K_m to make this conceivable.

D. AUTOCATALYSIS

Although ribulose diphosphate and phosphoenolpyruvate can be regenerated according to the processes outlined above, only the former is regenerated

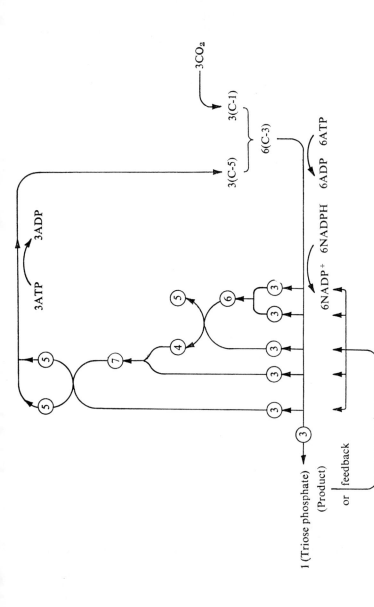

FIG. 5. The Benson-Calvin cycle as an autocatalytic sequence or breeder reaction. Represented in triplicate so that the condensation of three molecules of ribulose diphosphate with three molecules of CO_2 leads to the eventual formation of six molecules of triose phosphate. Normally five molecules of triose phosphate would then be rearranged in the reactions of the sugar phosphate "shuffle" to regenerate three molecules of CO_2-acceptor whilst the remaining triose phosphate would constitute net product. In an autocatalytic phase, it is assumed that the triose phosphate product could feed back into the cycle. Disregarding loss of triose phosphate in export from the chloroplast, etc., this would permit a theoretical doubling of the concentration of the acceptor for every fifteen molecules of CO_2 fixed. The circled numbers represent triose phosphates ③, erythrose-4-phosphate ④, pentose phosphates ◯ and sedoheptulose-1,7-diphosphate ⑦. Similarly C-1 is CO_2, C-3 is 3-phosphoglycerate and C-5 is ribulose-1,5-diphosphate. (After Walker, 1973, by courtesy of *New Phytologist*.)

in an autocatalytic sequence which is capable of growth. This is an important distinction which was occasionally lost sight of in the elucidation of C-4 photosynthesis and even led to statements which implied that phosphoenolpyruvate carboxylation might constitute an *alternative* to ribulose diphosphate carboxylation rather than an adjunct to it. In fact the Benson-Calvin cycle is a "breeder reaction" (in the sense that it can produce more ribulose diphosphate than it uses—see Fig. 5) whereas the regenerative process in C-4 photosynthesis (Fig. 8) can *replace* phosphoenolpyruvate but not *increase* it (except to the extent that CO_2, released by decarboxylation, can re-enter the Benson-Calvin cycle and yield phosphoenolpyruvate via

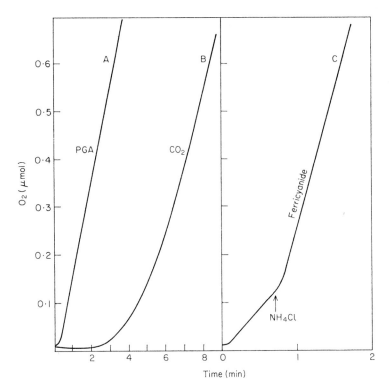

Fig. 6. Induction in chloroplasts. Curves A and C show virtual absence of induction of the Hill reaction with ferricyanide as an artificial oxidant and with 3-phosphoglycerate as the immediate precursor of the natural oxidant (1,3-diphosphoglycerate). The slight initial lag with ferricyanide is probably almost entirely attributable to the response time of the oxygen electrode. The slightly longer lag (note different time scale) with 3-phosphoglycerate is probably real and represents the time taken to build up 1,3-diphosphoglycerate. When CO_2 is the sole added substrate the lag is much longer because in order to attain maximal O_2 evolution all of the intermediates of the cycle will need to increase to their full steady state concentration for the prevailing conditions. This lag, therefore, reflects the autocatalytic increase in substrates represented by Fig. 5 which is an essential feature of a primary carboxylating mechanism. (From Walker, 1973, by courtesy of *New Phytologist*.)

phosphoglycerate). In this respect C-3 photosynthesis is self-sufficient whereas C-4 photosynthesis is dependent on the Benson-Calvin cycle and, therefore, is subsidiary rather than primary. That a primary carboxylating mechanism must have the capacity for growth and autocatalysis is self evident (Walker, 1973) but it is such an important feature of photosynthesis that it seems to justify some further clarification. This is quite simply that *all* metabolic processes other than that of primary carboxylation can continuously call on pre-elaborated carbon. Conversely, a primary carboxylating process must be independent and autocatalytic; either, in the short term, to accommodate advantageous changes in environment (such as increasing light intensity) or, in the long term, to permit growth of the organ or organism.

The Benson-Calvin cycle can, in theory, double its ribulose diphosphate concentration every 15 revolutions (Fig. 5) if all of its end product is repro- cessed for this purpose (Walker, 1973). Normally photosynthesis does not reach its maximal rate immediately after a prolonged dark interval but only following an initial lag or induction period (see e.g. Figs 6 and 7). This is believed to represent the period of autocatalysis during which intermediates of the cycle reach a high enough steady state level to maintain photosynthesis at the maximal rate permitted by external factors. Although this notion was

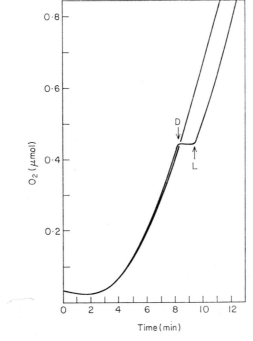

FIG. 7. Induction in chloroplasts. As for Fig. 6 showing absence of induction after a brief dark interval. Cycle intermediates having attained their maximal steady state concen- trations by autocatalysis are not as readily depleted in the dark in isolated chloroplasts as in some intact organisms. (After Walker, 1973, by courtesy of *New Phytologist*.)

first proposed many years ago by Osterhout and Haas (1918), it was not based on cyclic autocatalysis and it is perhaps surprising that the path of carbon in photosynthesis was not seen as a cyclic process prior to the work of Calvin and his colleagues. This is especially true in relation to the fact that photosynthesis was known to show a most unusual response to temperature under certain conditions, with Q_{10} values as high as 11 or 12 (Rabinowitch, 1956). None of the explanations put forward at the time embodied cyclic autocatalysis though (again with the benefit of hindsight) it may readily be seen that a cyclic sequence of reactions, each with normal temperature coefficients, would show an added response to increased temperature if all of the end product was fed back into the cycle and a ceiling had not already been imposed by other factors (Selwyn, 1966; Baldry *et al.*, 1966).

IV. C-4 Photosynthesis

A. the C-4 syndrome

Certain plants (including monocotyledons such as maize and sugar cane and dicotyledons such as *Amaranthus edulis*) share some or all of the features which constitute the C-4 syndrome (Downton, 1970). These include "Kranz

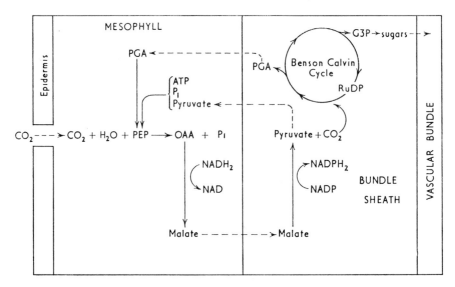

Fig. 8. C-4 photosynthesis. The above is one interpretation of results obtained with C-4 plants. CO_2 is believed to be fixed via phosphoenolpyruvate carboxylase in the outer (mesophyll) compartment and transported as malate to the inner (bundle sheath) compartment where, after decarboxylation, it is refixed in the Benson-Calvin cycle. Phosphoenolpyruvate is regenerated in a reaction sequence involving phosphate pyruvate dikinase and inorganic pyrophosphatase. An alternative interpretation would associate some Benson-Calvin cycle activity with both compartments and some direct fixation of CO_2 via this pathway.

type anatomy" (in which the photosynthetic tissues are arranged in concentric cylinders around the vascular bundles), low CO_2-compensation points, lack of ready photosaturation, absence of photorespiration, etc. When exposed briefly to $^{14}CO_2$ in the light these plants initially incorporate radioactivity into malate and aspartate rather than phosphoglycerate and sugar phosphates (see e.g. Hatch, 1970; Hatch and Slack, 1970). In longer term photosynthesis these latter compounds also become labelled. Similarly in "pulse chase" experiments (in which $^{14}CO_2$ is followed by $^{12}CO_2$) there is a quantitative transfer of label from malate and aspartate to Benson-Calvin cycle inter-mediates. This process of C-4 photosynthesis (so called because the first products are 4 carbon carboxylic acids rather than 3-phosphoglycerate and triose phosphates) is now held (Hatch, 1970) to be a sequence in which there is a preliminary fixation of CO_2 catalysed by phosphoenolpyruvate carbox-ylase in an outer (mesophyll) compartment. Malate and/or aspartate derived from this carboxylation step is transported (Fig. 8) into an inner (bundle sheath) compartment where they are decarboxylated. The CO_2 released is then refixed by ribulose diphosphate carboxylase (Hatch, 1970).

B. THE MALATE–CO_2 PUMP (Fig. 8)

This somewhat curious sequence of reactions is believed to constitute a malate-mediated CO_2 pump between the outer and inner compartments and it seems indisputable that C-4 plants fix nearly twice as much CO_2 per molecule of water lost by transpiration as do C-3 plants while still supporting very high rates of photosynthesis (Downton, 1970). Such a favourable ratio of CO_2 fixed to water lost would result from a very steep CO_2 gradient between the external atmosphere and the sub-stomatal spaces. It is arguable whether this steep gradient (also reflected by the low CO_2 compensation point) is brought about by the successful operation of the malate–CO_2 pump or by repression of photorespiratory CO_2 generation or by both.

C. PHOTORESPIRATION

C-3 plants "photorespire" particularly under conditions of intense illumi-nation, high temperatures and high oxygen tension (see e.g. Tolbert, 1970; Zelitch, 1971). Consequently they release CO_2 when illuminated in CO_2-free air and are unable to diminish the CO_2 in an enclosed space much below 50–100 ppm. C-4 plants do not release appreciable CO_2 in this way and have CO_2 compensation points nearer to 5 ppm. The source of CO_2 released by photorespiration is believed to be glycollate (Tolbert, 1970; Zelitch, 1971) which may be derived as phosphoglycollate from ribulose diphosphate in a reaction in which O_2 competes with CO_2 as shown below (Bowes and Ogren, 1972; Bowes et al., 1971; Andrews et al., 1973; Lorimer et al., 1973).
The lack of detectable photorespiration in C-4 plants has been attributed

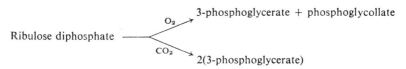

to several causes (see e.g. Tolbert, 1970) but re-fixation of CO_2 in the outer (mesophyll) compartment of photorespiratory CO_2 or suppression of phos-phoglycollate formation by the maintenance of high CO_2 in the inner (bundle sheath) compartment as a result of the operation of the malate–CO_2 pump must come high in the list of possible alternatives (Goldsworthy and Day, 1970; Zelitch, 1971). Such effects might well be additive.

D. C-4 PHOTOSYNTHESIS AS A SECONDARY PROCESS

Although it will be seen from Section IV A1 that C-4 photosynthesis involves a preliminary carboxylation catalysed by phosphoenolpyruvate carboxylase this is still a secondary process in the sense that it depends on the primary mechanism of the Benson-Calvin for its substrate and its ability to grow. C-4 photosynthesis is for this reason essentially an adjunct to the Benson-Calvin cycle which apparently increases its efficiency by decreasing concomitant water loss and photorespiration. It should also be noted that the schematic outline of C-4 photosynthesis shown in Fig. 8 represents only one extreme interpretation of the known facts and, as such, is in no way specifically endorsed by the present writer. Many would dispute the fact that Benson-Calvin cycle reactions occur only in the inner compartment (rather than in both) and some, for whom I have considerable sympathy, would hold that there is little about C-4 photosynthesis which could not be explained simply by high phosphoenolpyruvate carboxylase activity in the mesophyll and "Kranz-type anatomy" (see e.g. Coombs, 1973).

V. CRASSULACEAN ACID METABOLISM

A. GENERAL CHARACTERISTICS

A number of plants, mostly succulent in habit, exhibit what has been termed Crassulacean Acid Metabolism (see e.g. Thomas, 1949, 1951). This is characterized by a diurnal fluctuation in acid and an inverse fluctuation in starch and is associated with a massive dark fixation of carbon dioxide (for a review see Ranson and Thomas, 1960). Usually the acid which accumulates by night is malic and feeding experiments with $^{14}CO_2$ indicate that radio-activity in this compound is distributed between the two carboxyl groups in the ratio of one third to two thirds. Although there is still some lack of unanimity amongst workers in the field (see e.g. Sutton and Osmond, 1972) this is usually taken to indicate a double carboxylation (Bradbeer *et al.*,

1958). Starch is thought to be degraded by night in a modified version of the pentose phosphate pathway leading to the formation of ribulose diphosphate. This is then carboxylated to give 3-phosphoglycerate. Phosphoenolpyruvate derived from this source is carboxylated again to yield oxaloacetate (and hence aspartate and malate) with the "one third, two thirds" distribution of label shown in Fig. 9. The enzyme responsible for the second carboxylation is phosphoenolpyruvate carboxylase (Walker, 1962) and the aspartate and malate are derived by transamination or reduction of the first formed oxalo-acetate as in C-4 photosynthesis. CAM plants have a very high affinity for CO_2 and are capable of virtually eliminating free CO_2 from a finite enclosed space in the dark. It is evident, therefore, that if ribulose diphosphate carboxylase is concerned in the first stage of this process then it is not

Starch $\longrightarrow$

$$\text{Starch} \xrightarrow{\quad \textcircled{C}O_2 \quad}
\begin{array}{l}
CH_2OPO(OH)_2 \\
| \\
C{=}O \\
| \\
HCOH \\
| \\
HCOH \\
| \\
CH_2OPO(OH)_2
\end{array}
\; + H_2O \longrightarrow
\begin{array}{l}
CH_2OPO(OH)_2 \\
| \\
HCOH \\
| \\
\textcircled{C}OOH \\
\\
COOH \\
| \\
HCOH \\
| \\
CH_2OPO(OH)_2
\end{array}
\longrightarrow$$

ribulose-1,5-diphosphate   2(3-phosphoglycerate)

$$\begin{array}{l}
CH_2 \\
\| \\
COPO(OH)_2 \\
| \\
\textcircled{C}OOH \\
\\
COOH \\
| \\
COPO(OH)_2 \\
\| \\
CH_2
\end{array}
\xrightarrow[\substack{H_2O \\ H_2O}]{\substack{\textcircled{C}O_2 \\ \textcircled{C}O_2}}
\begin{array}{l}
4\,\textcircled{C}COH \\
| \\
3\,CH_2 \\
| \\
2\,C{=}O \\
| \\
1\,\textcircled{C}OOH \\
\\
1\,COOH \\
| \\
2\,C{=}O \\
| \\
3\,CH_2 \\
| \\
4\,\textcircled{C}OOH
\end{array}
\longrightarrow
\begin{array}{l}
{**}COOH \\
| \\
CH_2 \\
| \quad \text{(malate)} \\
HCOH \\
| \\
{*}COOH \\
\\
{*}COOH \\
| \\
HCNH_2 \\
| \quad \text{(aspartate)} \\
CH_2 \\
| \\
{**}COOH
\end{array}$$

2(phosphoenolpyruvate)   2(oxaloacetate)

FIG. 9. Proposed double carboxylation in Crassulacean Acid Metabolism. Ribulose diphosphate is believed to be derived from starch via a modified pentose phosphate pathway. Carboxylation by $^{14}CO_2$ would then give rise to two molecules of 3-phospho-glycerate (one labelled, one not). These, in turn, would yield phosphoenolpyruvate and a second carboxylation would give rise to two molecules of oxaloacetate (one labelled only in the β-carboxyl group, the other in both carboxyl groups). In total, this oxaloace-tate would show a "one third, two thirds" distribution of label between carbons 1 and 4. This labelling pattern would also appear in malate and aspartate derived from oxalo-acetate by reduction and transamination respectively. (Based on a proposal by Bradbeer *et al.*, 1958.)

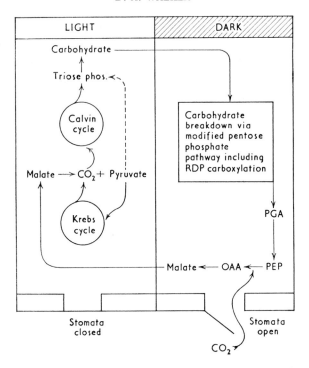

FIG. 10. Crassulacean Acid Metabolism. This figure represents two processes separated by time. In the dark there is a preliminary fixation of CO_2 into malate in which the carboxylation substrates are derived from starch synthesized in the previous light period. In the light the stomata close and the CO_2 released from decarboxylation of malate is refixed via the Benson-Calvin cycle. This produces the inversely related diurnal fluctuation in acid and starch which is characteristic of CAM plants and conserves water loss by restricting CO_2 uptake to the dark when the saturation deficit may be negligible.

switched off in the dark in the same way as in C-3 plants or, alternatively, that some mechanism exists by which CO_2/bicarbonate is maintained at higher concentrations at this carboxylation site than would exist in normal equilibrium with air. The overall process is outlined in Fig. 10 and it will be seen that dark fixation is believed to be capable of accounting for the total CO_2 uptake by the plant (see e.g. Neales *et al.*, 1968) and that, at least in some circumstances there is no stomatal exchange of CO_2 during the light. The stomata of CAM plants do not open by day in these circumstances and water loss from this cause is, therefore, avoided. During the day the acid which has accumulated by night undergoes β-decarboxylation (probably by the reverse of reaction 7) and the CO_2 so evolved is immediately refixed via the Benson-Calvin cycle. CAM plants often grow in semi-arid and even very arid habitats and despite the additional energy expenditure consequent upon this preliminary dark fixation the associated conservation of water is believed to afford a distinct ecological advantage.

B. CAM AS A SECONDARY PROCESS

Like C-4 photosynthesis CAM is a secondary process. In CAM the preliminary carboxylation is widely separated in time from the primary carboxylation of the Benson-Calvin cycle but again, as a continuing process it depends on the Benson-Calvin cycle for its substrate and its capacity for autocatalysis and growth. At present, CAM would be seen almost entirely as an adaptation to environment which permits plants to take up CO_2 by night and, therefore, diminishes stomatal water loss by day.

VI. CONCLUSION

The essential features of a primary carboxylating mechanism must include:

(1) a carboxylation with a favourable equilibrium position;
(2) a carboxylase with a high affinity for CO_2 or an associated CO_2 pump, or both;
(3) a process for the regeneration of the CO_2 acceptor;
(4) the ability for autocatalysis and growth.

Only the Benson-Calvin cycle possesses all of these properties and by definition, therefore, remains as the only known *primary* carboxylating mechanism. C-4 photosynthesis and CAM are adjuncts to the primary mechanism which enhance its efficiency and permit adaptation to otherwise unfavourable environments.

REFERENCES

Andrews, T. J., Lorimer, G. H. and Tolbert, N. E. (1973). *Biochemistry* **12**, 11.

Banks, B. E. C. and Vernon, C. A. (1970). *J. theor. Biol.* **29**, 301.

Baldry, C. W., Walker, D. A. and Bucke, C. (1966). *Biochem. J.* **101**, 641.

Bassham, J. A. and Calvin, M. (1957). "The Path of Carbon in Photosynthesis". Prentice Hall, Englewood Cliffs, N.J.

Bassham, J. A. and Jensen, R. G. (1966). *In* "Harvesting the Sun" (A. San Pietro, F. A. Greer and T. J. Army, eds). Proc. Int. Minerals Chem. Symp., Academic Press, New York, 1967.

Bassham, J. A. and Krause, G. H. (1969). *Biochim. biophys. Acta* **189**, 207.

Bassham, J. A., Sharpe, P. and Morris, I. (1968). *Biochim. biophys. Acta* **153**, 898.

Bowes, G. and Ogren, W. L. (1972). *J. biol. Chem.* **247**, 2171.

Bowes, G., Ogren, W. L. and Hageman, R. H. (1971). *Biochem. biophys. Res. Commun.* **45**, 716.

Bradbeer, J. W., Ranson, S. L. and Stiller, M. L. (1958). *Pl. Physiol.* **33**, 66.

Coombs, J. (1973). *Commentaries in Plant Science* **1**, 1.

Cooper, T. G., Filmer, D., Wichnick, M. and Lane, M. D. (1969). *J. biol. Chem.* **244**, 1081.

Cooper, T. G. and Wood, H. G. (1971). *J. biol. Chem.* **246**, 5488.

Downton, W. J. S. (1970). *In* "Photosynthesis and Photophosphorylation" (M. D. Hatch, C. B. Osmond and R. O. Slatyer, eds), p. 3. Wiley, New York, 1971.

Goldsworthy, A. and Day, P. R. (1970). *Nature, Lond.* **228**, 687.

Hatch, M. D. (1970). *In* "Photosynthesis and Photophosphorylation" (M. D. Hatch, C. B. Osmond and R. O. Slatyer, eds), p. 139. Wiley, New York, 1971.

Hatch, M. D. and Slack, C. R. (1970). *A. Rev. Pl. Physiol.* **2**, 141.

Herbert, D. (1951). *Symp. Soc. exp. Biol.* **5**, 52.

Jensen, R. G. (1971). *Biochim. biophys. Acta* **234**, 371.

Jensen, R. G. and Bassham, J. A. (1968). *Biochim. biophys. Acta* **153**, 227.

Krebs, H. A. (1954). *Bull. John Hopkins Hosp.* **95**, 19.

Lin, D. C. and Nobel, P. S. (1971). *Archs. Biochem. Biophys.* **145**, 622.

Lorimer, G. H., Andrews, T. J. and Tolbert, N. E. (1973). *Biochemistry* **12**, 18.

Neales, T. F., Patterson, A. A. and Hartney, V. J. (1968). *Nature, Lond.* **219**, 469.

Ochoa, S., Mehler, A. H. and Kornberg, A. (1947). *J. biol. Chem.* **167**, 871.

Osterhout, W. J. V. and Haas, A. R. C. (1918). *J. gen. Physiol.* **1**, 1.

Poincelot, R. P. (1972). *Biochim. biophys. Acta* **258**, 637.

Rabinowitch, E. I. (1956). "Photosynthesis and Related Processes". Vol. II, Pt. 2, p. 1313. Interscience, New York.

Ranson, S. L. and Thomas, M. (1960). *A. Rev. Pl. Physiol.* **11**, 81.

Selwyn, M. J. (1966). *Biochim. biophys. Acta* **126**, 214.

Stokes, D. M., Walker, D. A. and McCormick, A. V. (1971). *Proc. 2nd Int. Cong. Photosynthesis* (Stresa), p. 1779.

Sugiyama, T., Nakayama, N. and Akazawa, T. (1968). *Archs Biochem. Biophys.* **126**, 737.

Sugiyama, T. Matsumoto, C. and Akazawa, T. (1969). *Archs Biochem. Biophys.* **129**, 597.

Sutton, B. G. and Osmond, C. B. (1972). *Pl. Physiol.* **50**, 360.

Thomas, M. (1949). *New Phytol.* **48**, 390.

Thomas, M. (1951). *Symp. Soc. exp. Biol.* **5**, 72.

Tolbert, N. E. (1970). *In* "Photosynthesis and Photorespiration" (M. D. Hatch, C. B. Osmond and R. O. Slatyer, eds), p. 458. Wiley, New York, 1971.

Walker, D. A. (1957). *Biochem. J.* **67**, 73.

Walker, D. A. (1962). *Biol. Rev.* **37**, 215.

Walker, D. A. (1971). *Proc. 2nd Int. Cong. Photosynthesis* (Stresa), p. 1774.

Walker, D. A. (1973). *New Phytol.* **72**, 237.

Walker, D. A. and Brown, J. M. A. (1957). *Biochem. J.* **67**, 79.

Walker, D. A., McCormick, A. V. and Stokes, D. N. (1971). *Nature, Lond.* **233**, 346.

Weissbach, A., Horecker, B. L. and Hurwitz, J. (1956). *J. biol. Chem.* **218**, 795.

Werdan, K. and Heldt, H. W. (1972). *Biochim. biophys. Acta* **280**, 430.

Zelitch, I. (1971). "Photosynthesis, Photorespiration and Plant Productivity." Academic Press, New York and London.

CHAPTER 3

The Relationship between Gluconeogenesis and Carbohydrate Oxidation in Higher Plants

T. AP REES, S. M. THOMAS,* W. A. FULLER AND B. CHAPMAN

Botany School, University of Cambridge, England

I. INTRODUCTION

After photosynthesis, perhaps the most important instances of sugar formation in higher plants occur when storage material is mobilized. Sugar is formed from each of the major insoluble reserve products—polysaccharide, fat and protein—that are found in plants (Chen and Varner, 1969; Beevers, 1969; Stewart and Beevers, 1967). Most of this sugar is sucrose, although considerable amounts of stachyose are made in marrows (Thomas and ap Rees, 1972a). This formation of sugar results in the conversion of insoluble storage material into one or other of the small number of compounds used by plants in translocation. Thus the significance of the process is that it is part of the mechanism for mobilization of reserves and, as such, is of particular importance in germination. The formation of sucrose from polysaccharides is relatively straightforward when compared to the synthesis of sucrose from fat or protein. The latter processes involve a net synthesis of oxaloacetate from compounds that are intermediates of the tricarboxylic acid cycle, and a net synthesis of hexose monophosphate from an intermediate of glycolysis, phosphoenolpyruvate. The overall conversion achieved by these net syntheses may be called gluconeogenesis. The basic pathway of gluconeogenesis in plants has been established unequivocally by Beevers and his

* Present address: Rothamsted Experimental Station, Harpenden, Herts., England.

colleagues (Beevers, 1969) from studies of the fatty endosperm of germinating castor beans. No other tissue has been studied so thoroughly but there is adequate evidence that the pathway in Fig. 1 represents the general route in higher plants.

Examination of Fig. 1 shows that gluconeogenesis involves intermediates and enzymes that have counterparts in the tricarboxylic acid cycle and glycolysis. As the pathways of gluconeogenesis and carbohydrate oxidation are irreversible, the existence of common intermediates gives rise to the possibility of futile cycles in which the products of reactions in one pathway are promptly consumed as substrates for reactions in the other pathway. An unregulated futile cycle could drastically reduce net gluconeogenesis and

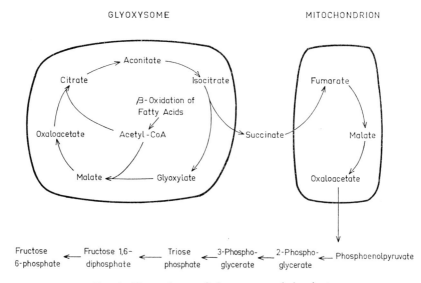

FIG. 1. The pathway of gluconeogenesis in plants.

could act as an ATPase at a crucial stage in plant development. In this chapter we discuss aspects of the relationship between gluconeogenesis and carbohydrate oxidation in plants. We pay specific attention to our own investigation of gluconeogenesis and glycolysis in the cotyledons of germinating marrows (*Cucurbita pepo* L. var. *medullosa* Alef.).

II. THE MAGNITUDE OF GLUCONEOGENESIS

It is not easy to measure gluconeogenesis in plants. In castor bean endosperm at the peak of gluconeogenesis, one gram of sugar accumulates for every gram of fat that disappears. During germination most of the fat is converted to sugars (Beevers, 1969). This means that loss via futile cycling must be small, and that rates of gluconeogenesis as high as the synthesis of 25 μmol hexose per hour per gram fresh weight must be attained (Cooper, 1971). This is a very substantial flux and it is ten times that of glycolysis by the same tissue under anaerobic conditions (Kobr and Beevers, 1971). Disappearance of storage fat is followed by senescence of the endosperm in castor bean. Thus biosynthetic processes in the endosperm, other than the synthesis of sugar, make relatively little demand on the storage fat. In this situation extensive conversion of storage fat to sucrose that is exported to the growing seedling is not unexpected.

The situation in storage tissues like the cotyledons of marrow and many other members of the Cucurbitaceae is more complicated and we have less idea of the magnitude of gluconeogenesis. These cotyledons do not senesce during germination but develop into photosynthetic tissues that persist in the young plant for some time. Here gluconeogenesis is accompanied by the differentiation of the fat-storing cells into normal photosynthetic cells. The whole process takes place rapidly and without detectable cell division (Gruber et al., 1970; Thomas and ap Rees, 1972a). In such cells gluconeo-genesis must coincide with very considerable demands for metabolic inter-mediates for biosyntheses. We do not know how these demands are met. We are particularly ignorant of the extent to which these demands are satisfied by gluconeogenesis. Breakdown of fat and storage protein could yield acetyl-CoA and amino acids that could be used directly for biosynthesis in the cotyledons. This mechanism would not involve gluconeogenesis but could significantly reduce the magnitude of gluconeogenesis by competing for the available fat and protein. At the other extreme, all the storage fat could be converted to sugar, and the needs for biosynthesis could be met by subsequent metabolism of some of this sugar via the conventional pathways of carbohydrate metabolism. A third possibility is that intermediates of gluconeogenesis could be diverted into the pathways of carbohydrate oxida-tion at certain specific and regulated points such as the reactions catalysed by citrate synthase and pyruvate kinase. Our general lack of knowledge in this field means that measurements of sugar accumulation in tissues like marrow cotyledons may not be accurate assessments of the magnitude of gluconeogenesis because they do not reveal whether any of the products of gluconeogenesis have been used in biosynthesis. Such measurements may indicate the minimum rate of gluconeogenesis. For marrow cotyledons we obtained a value of 3·8 μmol hexose per hour per gram fresh weight as an estimate of the minimum rate of gluconeogenesis. This is low compared to the value obtained for castor bean endosperm but the rate is still appreciable. Our estimate of the minimum rate of gluconeogenesis in marrow cotyledons is comparable to the maximum catalytic activity of phosphofructokinase and thus may be of the same order as the glycolytic capacity of the cotyledons (Thomas and ap Rees, 1972a,b).

The available evidence indicates that gluconeogenesis in fatty seedlings is a process of considerable magnitude. In considering the relationship between gluconeogenesis and carbohydrate oxidation, we should bear in mind the evidence that the rates of gluconeogenesis are comparable to or greatly in excess of the glycolytic capacity of gluconeogenic tissues.

III. THE GLYOXYLATE AND TRICARBOXYLIC ACID CYCLES

Acetyl-CoA, derived from storage fats, and intermediates of the tricarboxylic acid cycle, derived from storage protein, could be oxidized via the tricarboxylic acid cycle. In gluconeogenic tissues such oxidation must be restricted. In castor bean endosperm the restriction appears to be almost complete (Beevers, 1969). In tissues like the cotyledons of marrows the restriction may be partial. The distribution of label from [^{14}C]acetate reveals considerable synthesis of lipid, amino acids and protein at the peak of gluconeogenesis in marrow cotyledons (Thomas and ap Rees, 1972a). However, as discussed in the last section, we do not know precisely how the precursors for these syntheses are formed from the storage material.

The glyoxysome forms a major part of the mechanism that channels acetyl-CoA from fat into gluconeogenesis rather than into the tricarboxylic acid cycle. The location of the enzymes for β-oxidation and the glyoxylate cycle in the glyoxysome can ensure that the acetyl-CoA from fat does not enter the tricarboxylic acid cycle directly. This arrangement does not entirely prevent interaction between the glyoxylate and tricarboxylic acid cycles. The glyoxysome converts fatty acids to succinate. It seems clear that further metabolism of the succinate cannot be carried out by the glyoxysome and that conversion of this succinate to oxaloacetate is catalysed by a segment of the tricarboxylic acid cycle (Cooper and Beevers, 1969). For there to be quantitative conversion of succinate to sugar, the oxaloacetate formed in this segment of the tricarboxylic acid cycle must be converted to phosphoenolpyruvate and must not be metabolized by citrate synthase. In a sense the location of the enzymes for β-oxidation and the glyoxylate cycle in the glyoxysome does not prevent the intermingling of intermediates of gluconeogenesis and the tricarboxylic acid cycle but delays it until the decarboxylation steps of the cycle have been by-passed.

We do not know the mechanism that channels the oxaloacetate, that is formed in the mitochondria during gluconeogenesis, to phosphoenolpyruvate rather than into the tricarboxylic acid cycle. There are at least three fairly obvious possibilities. Firstly, diversion of the oxaloacetate to phosphoenolpyruvate would be favoured if gluconeogenic cells contained very little mitochondrial citrate synthase. This is almost certainly not the mechanism. Castor bean endosperm at peak gluconeogenesis is an excellent source of mitochondria that oxidize oxaloacetate rapidly (Beevers and Walker, 1956; Walker and Beevers, 1956). The available evidence indicates that the capacity of castor bean endosperm to oxidize intermediates of the tricarboxylic acid cycle rises markedly during the development of gluconeogenesis (Beevers and Walker, 1956). A second possibility is that citrate synthase in the mitochondria is inhibited during gluconeogenesis. Axelrod and Beevers (1972) have shown that the mitochondrial citrate synthase in castor bean endosperm is inhibited by ATP. In contrast citrate synthase from the glyoxysomes of the same tissue was not inhibited by ATP when assayed under similar conditions. Kobr and Beevers (1971) have provided evidence that the level of ATP in castor bean endosperm increases dramatically during gluconeogenesis. Treatments, that would be expected to reduce the level of ATP, produced effects on castor bean endosperm that could have been due to an increase in the amount of gluconeogenic oxaloacetate that entered the tricarboxylic acid cycle (Tanner and Beevers, 1965). Thus it seems likely that accumulation of ATP during gluconeogenesis may severely inhibit mitochondrial citrate synthase and thus direct oxaloacetate to phosphoenolpyruvate. It is clearly important to discover whether the concentration of ATP in the mitochondria during gluconeogenesis is high enough to achieve effective control of citrate synthase.

A third possible mechanism for restricting the metabolism of oxaloacetate by the tricarboxylic acid cycle is lack of acetyl-CoA in the mitochondria. This would deprive mitochondrial citrate synthase of an essential substrate. This could be achieved by the retention of acetyl-CoA by the glyoxysome and by a lack of appreciable acetyl-CoA synthesis by respiratory pathways. The extent to which the pathways of carbohydrate oxidation are inoperative during gluconeogenesis is difficult to assess. There is no doubt that both castor bean endosperm and marrow cotyledons, at peak gluconeogenesis, can oxidize sugars via glycolysis (Kobr and Beevers, 1971; Thomas and ap Rees, 1972b). In marrow cotyledons the capacity for glycolysis is comparable to the minimum estimate of the rate of gluconeogenesis. The efficiency of this third possible control mechanism will depend upon the relative rates of gluconeogenesis and carbohydrate oxidation and we cannot judge whether such a mechanism operates until we manage to measure the rates of gluconeogenesis and carbohydrate oxidation in the same tissue under natural conditions.

In conclusion, it seems likely that interaction between gluconeogenesis and the tricarboxylic acid cycle is governed partly by separation of the glyoxylate cycle from the decarboxylation reactions of the tricarboxylic acid cycle, party by inhibition of mitochondrial citrate synthase by ATP accumulating during gluconeogenesis, and, possibly, partly by a lack of acetyl-CoA in the mitochondria. It is important to appreciate that the last two mechanisms could vary considerably the proportion of oxaloacetate, produced in gluconeogenesis, that entered the tricarboxylic acid cycle. In this way some of the products of fat breakdown could be diverted from gluconeogenesis into the tricarboxylic acid cycle to support biosynthesis.

TABLE I

Development of gluconeogenesis in cotyledons of marrows germinated in the dark
(Thomas and ap Rees, 1972a)

Days germinated	Lipid (g/seedling)	Sugar (mg/seedling)	Metabolized $[2-^{14}C]$acetate recovered in sugar (%)	Fructose 1,6-diphosphatase[a] (nmol/min per cotyledon)
0	4.82	81	1·7	14·7 ± 1·4
2	4·66	45	1·1	40·5 ± 6·1
4	4·34	227	—	—
5	—	—	14·3	321 ± 24
6	2·02	895	—	—
8	2·20	894	11·0	332 ± 19

[a] Values are means ±S.E. of activities of six different extracts.

IV. GLUCONEOGENESIS AND GLYCOLYSIS

Part of the pathway of gluconeogenesis is, overall, the reversal of the changes catalysed by glycolysis. It is clear that quantitative conversion of phosphoenolpyruvate to sugars will require rigorous control of the steps in glycolysis that are catalysed by phosphofructokinase and pyruvate kinase. This question has received relatively little attention in plants and we concentrate upon data that we have obtained with the cotyledons of marrows germinated in the dark. Much of our work has been with 5-day-old seedlings. These are only slightly etiolated as under our conditions for germination the cotyledons do not emerge into the light until the seedlings are $4\frac{1}{2}$ days old. Therefore, we feel that our results do not represent an extreme physiological condition but are fairly close to the natural situation.

A. GLYCOLYTIC CAPACITY DURING GLUCONEOGENESIS

Phosphofructokinase and pyruvate kinase are the two enzymes, specific to glycolysis, that catalyse reactions that are irreversible *in vivo*. Interaction between gluconeogenesis and glycolysis could be regulated by maintenance of the maximum catalytic activities of these two enzymes at very low levels during gluconeogenesis. We have examined this possibility in respect of marrow cotyledons. The development of gluconeogenesis in the cotyledons of germinating marrows is shown by the data in Table I. Three criteria were used to assess gluconeogenesis. These were the changes in the contents of fat and sugar, the ability of the tissue to convert [2-^{14}C]acetate to sugar, and the activity of fructose 1,6-diphosphatase. No single type of measurement is conclusive, but the close agreement between all three types leaves little doubt that gluconeogenesis was slight or non-existent for the first two days of germination and that it then developed rapidly to reach a peak in 5- to 6-day-old seedlings.

Estimates of the activities of phosphofructokinase and pyruvate kinase in marrow cotyledons at different stages of germination are given in Table II.

TABLE II

Activities of phosphofructokinase and pyruvate kinase in extracts of cotyledons of marrows germinated in the dark (Thomas and ap Rees, 1972b)

Days germinated	Activity[a] (nmol/min per cotyledon)	
	Phosphofructokinase	Pyruvate kinase
0	3·54 ± 0·4	16·3 ± 1·2
2	4·75 ± 0·7	72·3 ± 12
5	9·43 ± 2·9	974 ± 124
8	11·80 ± 1·3	1485 ± 225

[a] Values are means ±S.E. of activities of six different extracts.

TABLE III

Activities of gluconeogenic and glycolytic enzymes in extracts of cotyledons of marrows germinated in the dark for 5 days

	Activity (nmol/min per cotyledon)[a]
Fructose 1,6-diphosphatase	321 ± 24
Phosphofructokinase	9·4 ± 3
Glyceraldehyde 3-phosphate dehydrogenase (NADH)	1475 ± 297
Phosphoglyceromutase	1978 ± 314
Enolase	2075 ± 254

[a] Data unpublished and from Thomas and ap Rees (1972a,b). Phosphoglyceromutase and enolase were assayed according to Shronk and Boxer (1964) and the other enzymes were assayed as described by Thomas and ap Rees (1972a,b). Values are means ±S.E. of activities of six different extracts.

These results indicate that the activities of the two specifically glycolytic enzymes rose significantly during the development of gluconeogenesis. The activity of phosphofructokinase, likely to be the key enzyme in determining glycolytic capacity, is low compared to that of the gluconeogenic enzymes (Table III). Nonetheless this activity is still appreciable when compared to our minimum estimates of the rate of gluconeogenesis and to the activity of phosphofructokinase in rapidly respiring non-gluconeogenic plant tissues (Fowler and ap Rees, 1970; Ricardo and ap Rees, 1972).

The changes in the activities of phosphofructokinase and pyruvate kinase indicate that the glycolytic capacity of marrow cotyledons increases during the development of gluconeogenesis. Additional evidence for this view is provided by comparison of the patterns of $^{14}CO_2$ production from specifically labelled glucose supplied to cotyledons from 2- and 5-day-old seedlings (Table IV).

TABLE IV

Release of $^{14}CO_2$ from specifically labelled glucose by cotyledons of marrows germinated in the dark (Thomas and ap Rees, 1972b)

Days germinated	Position of ^{14}C in [^{14}C]glucose	% of added [^{14}C]glucose recovered as $^{14}CO_2$ in:				
		1	2	3	4	6 h[a]
2	1	0·05	0·34	0·83	1·26	2·20
	2	0·01	0·09	0·26	0·47	1·10
	3,4	0·10	0·70	2·13	3·58	6·83
	6	0·02	0·12	0·31	0·54	1·11
5	1	0·08	0·32	0·64	1·05	1·86
	2	0·05	0·20	0·44	0·77	1·51
	3,4	0·36	2·52	5·53	8·78	15·80
	6	0·05	0·20	0·39	0·60	1·06

[a] Hours from addition of [^{14}C]glucose to samples. All values are means of triplicate samples.

These patterns show that cotyledons of both ages released $^{14}CO_2$ from [3,4-^{14}C]glucose far more readily than from [1-^{14}C]-, [2-^{14}C]- or [6-^{14}C]-glucose. Although glucose carbons 3 and 4 can be released via the pentose phosphate pathway, such release would not be expected to exceed that from carbon 1. The relatively high yields from carbons 3 and 4 indicate that the cotyledons of both ages metabolized most of the exogenous glucose via glycolysis. The only significant difference between the cotyledons of the two ages is that the yields from carbons 3 and 4, but not from any other carbon, increased sharply between days 2 and 5. This indicates that cotyledons from 5-day-old marrows metabolized more of the added glucose via glycolysis than did cotyledons from 2-day-old marrows. This change coincided with the increase in enzyme activities (Table II) and with the development of gluconeo-genesis. In considering these results it is important to bear in mind that there is evidence that the glycolytic capacity of castor bean endosperm increases during the development of gluconeogenesis (Kobr and Beevers, 1971).

The few data that are available indicate that maintenance of the maximum catalytic activities of the two specifically glycolytic enzymes at an abnormally low level is not a prime means of controlling interactions between gluconeo-genesis and glycolysis in plants. Indeed the data indicate that glycolytic capacity increases during the development of gluconeogenesis.

B. LOCATION OF GLUCONEOGENESIS FROM PHOSPHOENOLPYRUVATE

Interactions between gluconeogenesis and glycolysis could be controlled by location of the gluconeogenic and glycolytic enzymes in different compart-ments. Kobr and Beevers (1971) have suggested that the gluconeogenic steps from phosphoenolpyruvate to sugar might be located in the proplastids, thus allowing gluconeogenesis and glycolysis to proceed independently. Subse-quently we made a similar suggestion to explain results that we obtained with marrow cotyledons (Thomas and ap Rees, 1972b). We have begun an inves-tigation of the role of compartmentation in regulating the relationship between gluconeogenesis and carbohydrate oxidation and we now report our initial findings. All the work has been carried out with cotyledons from marrows that have germinated in the dark for 5 days.

First, we considered whether the gluconeogenic enzymes are in different cells from the glycolytic enzymes. Apart from a little vascular tissue, 5-day-old cotyledons can be divided into a mass of mesophyll cells and two or three layers of pallisade cells (Fig. 2). We dissected mesophyll cells free from pallisade cells until we obtained samples of the two types of cells that were large enough to assay for characteristic enzymes of gluconeogenesis and glycolysis (Table V). The results show that both types of cells contain both gluconeogenic and glycolytic enzymes. We suggest that the gluconeogenic and glycolytic enzymes in marrow cotyledons are almost certainly not separ-ated by compartmentation in different types of cells.

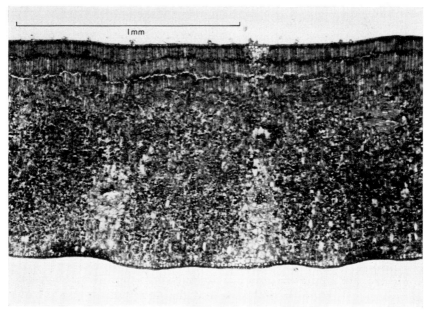

Fig. 2. Transverse section of cotyledon of marrow germinated in the dark for 5 days. The section was stained with safranin and fast green.

TABLE V

Intercellular distribution of enzymes of gluconeogenesis and glycolysis in cotyledons of marrows germinated in the dark for 5 days[a]

Enzyme	Activity[b] (μmol/min per g fresh wt)	
	Pallisade cells	Mesophyll cells
Isocitrate lyase	3·17 ± 0·29	4·47 ± 0·19
Fructose 1,6-diphosphatase	2·04 ± 0·20	2·09 ± 0·16
Phosphoglyceromutase	13·43 ± 1·31	10·59 ± 0·59
Phosphofructokinase	0·27 ± 0·06	0·43 ± 0·02
Alcohol dehydrogenase	1·73 ± 0·29	2·18 ± 0·29

[a] The pallisade cells were dissected free from the mesophyll cells and samples (0·6 g fresh wt) extracted and assayed. Fructose 1,6-diphosphatase and phosphofructokinase were assayed, as described by Thomas and ap Rees (1972a,b). Isocitrate lyase was assayed according to Cooper and Beevers (1969), phosphoglyceromutase according to Shronk and Boxer (1964), and alcohol dehydrogenase according to Racker (1955).

[b] Values are means ±S.E. of activities of four different extracts. (Unpublished data.)

Next we investigated whether the gluconeogenic enzymes are separated from those of glycolysis by intracellular compartmentation. Examination of the cotyledons of 5-day-old marrows showed mitochondria, glyoxysomes, and proplastids as possible compartments for the segregation of the gluconeogenic enzymes from the glycolytic enzymes (Fig. 3). In testing the hypothesis that gluconeogenesis from phosphoenolpyruvate occurs within a compartment that is separate from the glycolytic enzymes, it is important

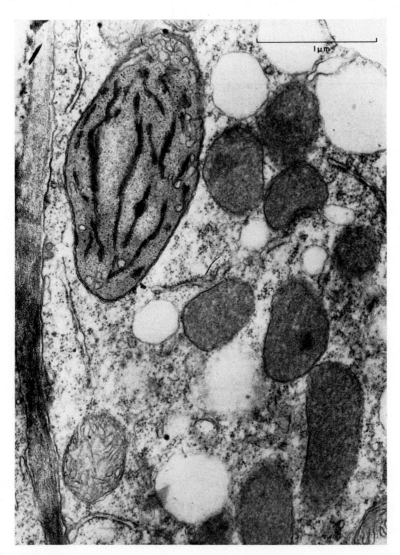

FIG. 3. Electron micrograph of cotyledon of marrow germinated in the dark for 5 days. Slices (1 mm thick) were fixed in glutaraldehyde and osmium tetroxide, embedded, sectioned, and then stained with lead citrate.

to bear in mind that there are proplastids in marrow cotyledons. These proplastids may contain significant activities of the enzymes involved in the photosynthetic conversion of 3-phosphoglycerate to hexose monophosphate. Although a gluconeogenic compartment must contain these enzymes, their presence cannot be regarded as a unique feature of such a compartment. Compartmentation of the gluconeogenic enzymes, resulting in their separation from the glycolytic enzymes, would only be revealed by the distribution of enolase and phosphoglyceromutase on the one hand, and the uniquely glycolytic enzymes, phosphofructokinase, pyruvate kinase, and alcohol dehydrogenase, on the other hand.

We have used the techniques of cell fractionation to assess the distribution of gluconeogenic enzymes in the cotyledons of 5-day-old marrows. In these experiments we took the following precautions. First, the cell fractions were examined with the electron microscope so that the structure of the isolated organelles could be compared to that seen in intact cells. Second, we measured the extent to which characteristic marker enzymes were retained within the isolated organelles. Third, the activities of enzymes recovered in isolated organelles were related to the total activities present in the unfractionated extracts. Finally, we emphasize the fact that the cotyledons of 5-day-old marrows contain high activities of the enzymes that we studied (Table III). As gluconeogenesis is one of the major physiological processes occurring in these cotyledons, a gluconeogenic compartment might be expected to contain substantial activities of the appropriate enzymes.

Mitochondria and glyoxysomes were isolated by density gradient centrifugation of material that sedimented at 10 800g (Fig. 4). We obtained two separate bands of mitochondria. In the upper band most of the mitochondria had the condensed configuration but in the lower band most of them were

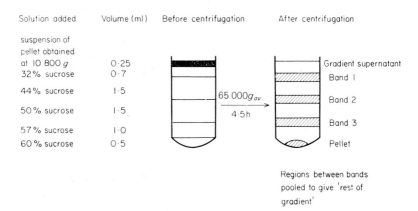

FIG. 4. Isolation of mitochondria and glyoxysomes from cotyledons of marrows germinated in the dark for 5 days. Samples of cotyledons were homogenized as described in Table VI to give material that sedimented at 10 800g. This material was fractionated on a discontinuous sucrose gradient as shown.

in the orthodox configuration (Fig. 5). Both bands were fairly free of con-
taminants. The glyoxysomes, contaminated with some debris, were recovered
in the pellet (Fig. 5). The enzyme activities of these fractions are shown in
Table VI. The distribution of the two marker enzymes for mitochondria,
fumarase and cytochrome oxidase, indicates that very few of the mitochondria
were broken in preparing the homogenates and that there was relatively

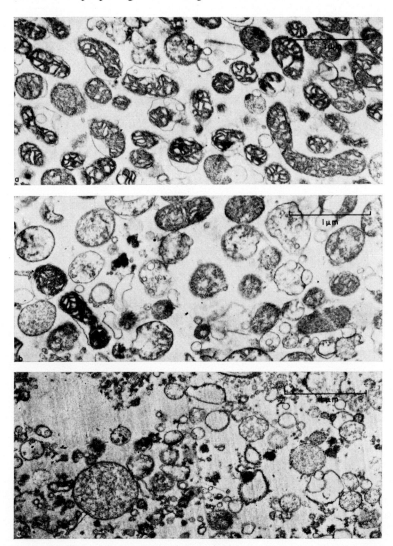

FIG. 5. Electron micrographs of mitochondrial and glyoxysomal fractions isolated, as
shown in Fig. 4, from cotyledons of marrows germinated in the dark for 5 days: a,
mitochondria (band 2); b, mitochondria (band 3); c, glyoxysomes (pellet). Fractions
were isolated from the gradient, centrifuged at 10 800g, and the sediments were then
processed as described under Fig. 3.

TABLE VI

Intracellular distribution of gluconeogenic enzymes in cotyledons of marrows germinated in the dark for 5 days: activities in mitochondria and glyoxysomes[a]

Cell fraction	Enzyme activity per fraction as percentage of total activity in unfractionated homogenate				
	Fructose 1,6-diphosphatase	Phosphogly-ceromutase	Isocitrate lyase	Fumarase	Cytochrome oxidase
Pellet from unfractionated homogenate at 10 800$g_{(av)}$	21	4	38	92	78
Fractions obtained from above pellet by centrifuging on a gradient as in Fig. 4					
Gradient supernatant	5·9	3·1	5·7	11	3
Band 1	2·7	0·02	0·6	n.d.[b]	4
Band 2 (mitochondria)	0·8	n.d.	0·6	31	20
Band 3 (mitochondria)	0·7	n.d.	0·6	31	35
Pellet (glyoxysomes)	8·0	0·4	26	n.d.	5
Rest of gradient	2·0	n.d.	6·0	22	12

[a] Samples of cotyledons (20 g) were extracted as described by Harris and Northcote (1971) in 20 ml 0·167 M glycylglycine (pH 7·4) that was 0·4 M with respect to sucrose. The extract was filtered through two layers of muslin and centrifuged at 500$g_{(av)}$. The supernatant is the unfractionated homogenate. This was centrifuged at 10 800$g_{(av)}$ and the pellet was fractionated on a sucrose gradient (Fig. 4). Enzymes were assayed as in Table V except that fumarase was assayed according to Cooper and Beevers (1969) and cytochrome oxidase according to Fritz and Beevers (1955). (Unpublished data.)

[b] n.d.; activity not detected.

little breakage during the density gradient centrifugation. The distribution of isocitrate lyase shows that a substantial proportion of the glyoxysomes was recovered intact but that breakage was greater than occurred with mitochondria. The central feature of the results in Table VI is that we found little or no phosphoglyceromutase activity in either the mitochondrial or glyoxysomal fractions. The electron micrographs and the retention of marker enzymes by the isolated organelles indicate that the absence of phospho-glyceromutase from mitochondria and glyoxysomes is not an artefact of extraction. These results suggest that the enzymes that catalyse the conversion of phosphoenolpyruvate to hexose monophosphate in marrow cotyledons are not located in either the mitochondria or the glyoxysomes.

The technique that we used to prepare mitochondria and glyoxysomes did not yield a proplastid fraction. There were some proplastids in the material that sedimented from the crude homogenate at $10\,800g$ but these were broken in the density gradient centrifugation. We tested a variety of methods for preparing proplastids. In general we found that proplastids from marrow cotyledons were very fragile and were particularly susceptible to centrifugation at speeds above $5000g$ and to re-suspension after centrifigation. We finally adopted a procedure that involved the use of a large column of "Sephadex" (Fig. 6). We stress that the fractions eluted from this column were centrifuged

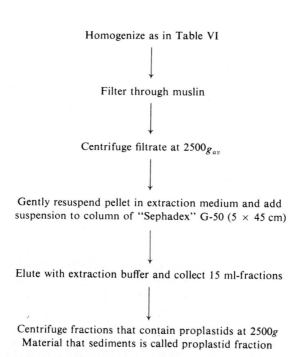

Homogenize as in Table VI

Filter through muslin

Centrifuge filtrate at $2500g_{av}$

Gently resuspend pellet in extraction medium and add suspension to column of "Sephadex" G-50 (5 × 45 cm)

Elute with extraction buffer and collect 15 ml-fractions

Centrifuge fractions that contain proplastids at $2500g$
Material that sediments is called proplastid fraction

FIG. 6. Preparation of proplastids from cotyledons of marrows germinated in the dark for 5 days.

at 2500*g* and that all our assays refer only to the material that sedimented. We used fructose 1,6-diphosphatase as a marker for proplastids. Elution of the column always gave a marked peak of fructose 1,6-diphosphatase activity that coincided with a peak of protein (Fig. 7). The precise position of this peak varied by a fraction or so with different columns but the coincidence between protein and fructose 1,6-diphosphatase was always found. Examination of the fractions that had fructose 1,6-diphosphatase activity showed that each of them contained considerable numbers of proplastids. Some of these

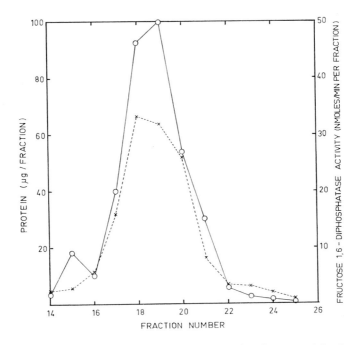

FIG. 7. Protein content (0——0) and fructose 1,6-diphosphatase activity (× – – – ×) of proplastid fractions obtained, as shown in Fig. 5, from cotyledons of marrows germinated in the dark for 5 days.

proplastids were broken but most appeared similar to those seen in intact cells (Fig. 8).

The enzyme activities of the proplastid fractions are given in Table VII. In these experiments all the fractions that contained proplastids were amalgamated and the total activity of the amalgamated fractions was measured and is given as activity in the proplastid fraction. We make two major points about the data in Table VII. First, there was a concentration of ribulose diphosphate carboxylase in the proplastid fractions. These data and the electron micrographs (Fig. 8) strongly indicate that our preparations contained a significant concentration of intact proplastids. Our data also show that many of the proplastids of the cells that were homogenized were broken during our preparative procedures. If we assume that all of the ribulose diphosphate carboxylase is confined to proplastids *in vivo*, then we can argue that we recovered 7% of the proplastids present in the cells.

The second feature of the data in Table VII that we stress is that although both enolase and phosphoglyceromutase were detected in the proplastid fractions, the percentages of their total activity that were recovered in these

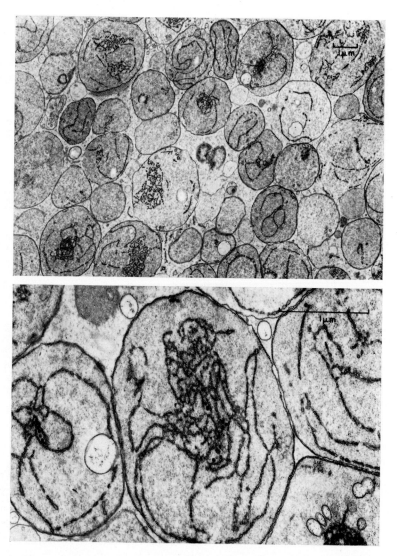

FIG. 8. Electron micrographs of proplastid fractions shown in Fig. 7. The sedimented proplastids were processed as described under Fig. 3.

TABLE VII

Intracellular distribution of gluconeogenic enzymes in cotyledons of marrows germinated in the dark for 5 days: activities in proplastid fractions

Enzyme	Enzyme activity per fraction as percentage of total activity in unfractionated homogenate[a]	
	Pellet from unfractionated homogenate at 2500g	Proplastid fraction
Ribulose 1,5-diphosphate carboxylase	33	7
Fructose 1,6-diphosphatase	19	1·1
Phosphoglyceromutase	1·6	0·05
Enolase	2·0	0·18

[a] Tissue homogenized as in Table VI and fractionated as in Fig. 6. Enzymes assayed as in Table III. Ribulose 1,5-diphosphate carboxylase assayed spectrophotometrically according to Andrews and Hatch (1971). Values are means of at least five different experiments. (Unpublished data.)

fractions are minute. If the argument that only 7% of the proplastids were recovered intact is accepted, then it can be maintained that the activities of enolase and phosphoglyceromutase in the proplastid fractions are only 7% of their true values. Even if this correction is applied the percentages of enolase and phosphoglyceromutase in the proplastid fractions can only be increased to 0·7% and 2·6%, respectively. We think that even these corrected values are far too small to support the view that the gluconeogenic enzymes are located in the proplastids.

We do not wish to come to any firm conclusion on the basis of the above results. The study is not yet complete and caution is necessary in the interpretation of data obtained by the fractionation of plant cells. Nonetheless, all our results are consistent with the view that gluconeogenesis from phosphoenolpyruvate in marrow cotyledons occurs in the cytoplasmic phase of the cells and not in mitochondria, glyoxysomes, or proplastids. Thus, at present, there is no direct evidence that intracellular compartmentation, involving membrane-bounded organelles, is a major means of regulating the interaction of gluconeogenesis and glycolysis.

The above considerations imply that interactions between gluconeogenesis and glycolysis may be regulated primarily by fine control of phosphofructokinase and pyruvate kinase *in vivo*. We know that phosphofructokinase from plants shows regulatory properties and is inhibited by ATP (Kelly and Turner, 1970). There have been fewer studies of pyruvate kinase from plants but the recent paper by Duggleby and Dennis (1973) is of particular importance in the present context. This paper reports that pyruvate kinase from ungerminated seeds of cotton showed regulatory properties in that ATP, UTP, citrate and malate were inhibitory, whilst AMP, GMP and fumarate functioned as activators. This indicates that pyruvate kinase could be inhibited in cells in which the ATP level is high and the tricarboxylic acid cycle is restricted. Such conditions may well arise during gluconeogenesis in plants. Thus it seems likely that the onset of gluconeogenesis in plants leads to conditions, in particular a rapid increase in ATP (Kobr and Beevers, 1971), that result in inhibition of the two specifically glycolytic enzymes. Support for this view is provided by the changes in substrate levels that occur when gluconeogenic endosperm of castor beans is deprived of oxygen (Kobr and Beevers, 1971). These changes indicate that glycolysis during gluconeogenesis is limited at the steps that are catalysed by phosphofructokinase and pyruvate kinase.

In conclusion, we consider the relative roles of the different mechanisms for regulating the interaction between gluconeogenesis and carbohydrate oxidation. There is no evidence for regulation by the maintenance, at an abnormally low level, of the maximum catalytic activities of the enzymes specific to carbohydrate oxidation. In each case that has been examined the evidence indicates that the activity of such enzymes increases during the development of gluconeogenesis. This suggests that during gluconeogenesis there is a need for metabolism through at least some segments of the pathways of carbohydrate oxidation in the direction opposite to that of gluconeogenesis. This need could be related to synthesis of intermediates required in the biosynthesis of cellular material, and it may be much greater in tissues like marrow cotyledons than in castor bean endosperm. Compartmentation is clearly of considerable importance but its role may be limited to the steps that precede the formation of oxaloacetate. We still know very little about the regulation of the steps between oxaloacetate and sugar. The data that are available point to a major, if not predominant, role of fine control.

ACKNOWLEDGEMENTS

We thank Miss Sylvia Bishop for her expert help. S. M. Thomas thanks the Science Research Council for a Research Studentship.

REFERENCES

Andrews, T. J. and Hatch, M. D. (1971). *Phytochemistry* **10**, 9.
Axelrod, B. and Beevers, H. (1972). *Biochim. biophys. Acta* **256**, 175.
Beevers, H. (1969). *Ann. N.Y. Acad. Sci.* **168**, 313.
Beevers, H. and Walker, D. A. (1956). *Biochem. J.* **62**, 114.
Chen, S. S. C. and Varner, J. E. (1969). *Pl. Physiol.* **44**, 770.
Cooper, T. G. (1971). *J. biol. Chem.* **246**, 3451.
Cooper, T. G. and Beevers, H. (1969). *J. biol. Chem.* **244**, 3507.
Duggleby, R. G. and Dennis, D. T. (1973). *Archs Biochem. Biophys.* **155**, 270.
Fowler, M. W. and ap Rees, T. (1970). *Biochim. biophys. Acta* **201**, 33.

Fritz, G. and Beevers, H. (1955). *Pl. Physiol.* **30**, 309.

Gruber, P. J., Trelease, R. N., Becker, W. M. and Newcomb, E. H. (1970). *Planta* **93**, 269.

Harris, P. J. and Northcote, D. H. (1971). *Biochim. biophys. Acta* **237**, 56.

Kelly, G. J. and Turner, J. F. (1970). *Biochim. biophys. Acta* **208**, 360.

Kobr, M. J. and Beevers, H. (1971). *Pl. Physiol.* **47**, 48.

Racker, E. (1955). *In* "Methods in Enzymology" (S. P. Colowick and N. O. Kaplan, eds), Vol. 1, p. 500. Academic Press, New York and London.

Ricardo, C. P. P. and ap Rees, T. (1972). *Phytochemistry* **11**, 623.

Shronk, C. E. and Boxer, G. E. (1964). *Cancer Res.* **24**, 709.

Stewart, C. R. and Beevers, H. (1967). *Pl. Physiol.* **42**, 1587.

Tanner, W. and Beevers, H. (1965). *Z. Pflanzenphysiol.* **53**, 126.

Thomas, S. M. and ap Rees, T. (1972a). *Phytochemistry* **11**, 2177.

Thomas, S. M. and ap Rees, T. (1972b). *Phytochemistry* **11**, 2187.

Walker, D. A. and Beevers, H. (1956). *Biochem. J.* **62**, 120.

CHAPTER 4

Plant Polyols

B. E. STACEY

*Department of Chemistry, Sir John Cass School of Science and Technology,
City of London Polytechnic, London, England*

I. Introduction

The term "polyol" refers to a compound containing three or more hydroxyl groups and as such includes glycerol, the alditols and the inositols. For the purpose of this review, however, I would like to restrict the discussion to the *alditols* which are best considered as reduction products of monosaccharides. Thus, reduction of D-arabinose gives D-arabinitol (classified as a pentitol since it contains five carbon atoms) while D-glucitol (a hexitol, sometimes called sorbitol) is derived from D-glucose.

II. Occurrence of Alditols in Plants

A general indication of the occurrence of alditols in plants is given in Table I. Mannitol is the most common of the alditols in both fungi and green plants while arabinitol is relatively common in fungi and lichens: Lindberg *et al.* (1953) found arabinitol in 55 out of 60 species of lichens they examined. Alditols are one of the principal soluble carbohydrate components of many fungi and lichens. Most of the algae in which alditols have been reported are marine algae and mannitol appears to be almost universal in

TABLE I

Occurrence of alditols in plants

	Fungi	Algae	Lichens	Higher plants
Erythritol	+	+	+	+
D-Threitol	+	−	−	−
D-Arabinitol	+	−	+	−
L-Arabinitol	+	−	−	−
Ribitol	−	+	+	+
Xylitol	+	−	−	−
D-Mannitol	+	+	+	+
D-Glucitol	+	+	−	+
Galactitol	+	+	−	+
Allitol	−	−	−	+
L-Iditol	−	−	−	+
Volemitol (D-*Glycero*-D-*manno*-heptitol)	+	+	+	+
Perseitol (D-*Glycero*-D-*galacto*-heptitol)	−	−	−	+
β-Sedoheptitol (D-*Glycero*-D-*gluco*-heptitol)	−	−	−	+
Meso-Glycero-ido-heptitol	+	−	−	−
D-*Glycero*-D-*ido*-heptitol	+	−	−	−
D-*Erythro*-D-*galacto*-octitol	−	−	−	+

the Phaeophyta (Lewis and Smith, 1967a). Volemitol (D-*glycero*-D-*manno*-heptitol) is present in *Pelvetia canaliculata* (Lindberg and Paju, 1954; Quillet, 1957). In higher plants, although mannitol is the most widely distributed of the alditols, some of the others are characteristic of particular families, e.g. D-glucitol in the Rosaceae and galactitol in the Celastraceae (Plouvier, 1963). Free ribitol is present in *Adonis* and *Bupleurum* while the remaining alditols that have been found in angiosperms are restricted as far as is known to either single genera or even single species, e.g. allitol in *Itea* spp. It is interesting to note that the occurrence of mannitol in fruits is rare but D-glucitol occurs in many, notably those of the Rosaceae. Two unusual alditols, D-*glycero*-D-*galacto*-heptitol (perseitol) and D-*erythro*-D-*galacto*-octitol, are constituents of the avocado pear.

Adequate and simple methods are at present available for the detection and estimation of alditols in plant tissues whereas older techniques frequently failed to reveal the presence of these compounds especially when accompanied by the corresponding monosaccharides. Modern methods of analysis include paper chromatography, paper electrophoresis, thin layer chromatography and gas liquid chromatography and these have been reviewed by Lewis and Smith (1967b). Table II gives an indication of the kind of separations that can be achieved using paper chromatography.

Polyol levels appear to be governed by different factors in different groups

TABLE II

Paper chromatography of sugars and alditols (Solvent: methyl ethyl ketone/acetic acid/water saturated with boric acid (9:1:1, v/v)) (see Lewis and Smith, 1967b)

	$R_{mannitol}$
Trehalose	4
Sucrose	12
Glucose	39
Galactose	39
Mannose	59
Fructose	65
Arabinose	78
Galactitol	96
Mannitol	100
Xylose	106
Glucitol	116
Arabinitol	152
Xylitol	156
Ribitol	165
Erythritol	193
Ribose	251

of plants. Black (1950), for example, investigated the seasonal variation in mannitol content of fronds of *Laminaria cloustoni*. The mannitol content of fronds collected near Cullipool in Scotland, was a minimum of about 6–8% dry weight in March with a maximum of 23–26% in the summer; an even wider fluctuation was noted with fronds collected in the Orkneys, with minima of about 8% and maxima of the order of 35%. In the latter location, changes with depths were noticed, maximum values being at 6–10 m. The seasonal variation of mannitol, laminarin and alginic acid in economically important seaweeds has been reviewed by Boney (1965); general trends in mannitol content are similar to the examples quoted above with maxima during late spring or early summer.

In higher plants, seasonal variations of mannitol have been found to occur in leaves of the olive and *Gardenia* sp. In the former, values varied from 6·1% dry weight in January to 2·1% in June rising again in the latter part of the year to 4·76% in December (Nuccorini, 1930). In *Gardenia*, mannitol was present only between September and May rising to a pronounced peak of 8% dry weight in February (Asai, 1937). The high winter levels of mannitol in these species might suggest the polyol could function osmotically in the prevention of frost damage, but this is not supported by the observations of Sakai (1961) who could find no consistent correlation between frost hardiness and polyol content in several woody species and indeed a contrasting situation occurs in apple, where maximum polyol content occurs in the summer.

III. Accumulation of Alditols during Photosynthesis

Alditols are frequently found to be major products of photosynthesis.

Bidwell (1958) has studied the products of photosynthesis of 14 species of marine algae in $^{14}CO_2$ and a selection of his results are presented in Table III: they show an obvious taxonomic relationship in that the main product was mannitol in the brown algae, glycerol glycosides in the red and sucrose in green algae.

When lichen tissues are incubated in solutions of $NaH^{14}CO_3$ in light most of the ^{14}C usually accumulates in alditols. Thus, in *Peltigera polydactyla*, over 60% of the fixed ^{14}C had accumulated in mannitol after 45 min (Drew, 1966), while in *Xanthoria aureola* over 90% of the ^{14}C was found in polyols (Bednar and Smith, 1966).

Accumulation of ^{14}C in alditols has also been found to occur in various higher plants during photosynthesis in $^{14}CO_2$, for example, allitol in *Itea* leaves (Hough and Stacey, 1966a), D-glucitol in plum leaves (Anderson *et al.*,

TABLE III

Photosynthesis of marine algae in $^{14}CO_2$ (4–6 h) (Bidwell, 1958)

	^{14}C % of soluble activity		
Species	Sucrose	Mannitol	Floridoside[a]
Fucus versiculosus	—	89	—
Alaria exulenta		88	—
Laminaria digitata		93	
Coralline officinalis	—	—	46
Chondrus crispus			65
Rhodymenia palmata	—	—	92
Cladophora sp.	73	2·4	4·8
Ulva lactuca	59	—	—

[a] 2-O-α-D-galactopyranosylglycerol.

1961) and mannitol in the leaves of lilac, white ash and celery (Trip *et al.*, 1963). Incorporation of ^{14}C into allitol in *Itea* leaves was rapid, accounting for 21% of the ^{14}C in alcohol-soluble materials after the first hour of photosynthesis and labelling increased rapidly to over 50% after 8 h, followed by a more gradual rise to an equilibrium value of over 60% after 24 h. The allitol content of the leaves showed a rapid increase during the first 2 h of photosynthesis (2·5–5% dry weight) followed by a more gradual increase to 9% after 48 h (Hough and Stacey, 1966b). This is in contrast to the situation with D-glucitol in plum leaves where although there was a rapid incorporation of ^{14}C there was no net synthesis of the alditol during photosynthesis. It was estimated that after photosynthesis for only 3 h, 15% of the leaf glucitol was newly synthesized from labelled precursors corresponding to the synthesis and utilization of about 2·0 mg of glucitol by each leaf. Clearly, there was a rapid turnover of the glucitol pool in equilibrium with primary products of photosynthesis whereas in the case of allitol it appears that this hexitol acts in a storage capacity.

IV. ALDITOLS AS ACTIVE METABOLITES

Various alditols have been shown to be active metabolites in a variety of plants. For example, when the lichen *Peltigera polydactyla* is starved in the dark at room temperature there is a fall in tissue mannitol content of 40% during the first 48 h, but thereafter the decline is much slower (Drew, 1966). Drew labelled the internal pool of mannitol in *P. polydactyla* by feeding trace amounts of ^{14}C-mannitol and then studied the fate of ^{14}C in tissues during 6 days' starvation in the dark at room temperatures; the results strongly suggest that mannitol is the major respiratory substrate of *P. polydactyla* and it was calculated that the rate of loss of carbon from mannitol and its appearance in $^{14}CO_2$ are very similar during the first 2 days of starvation.

Mention has been made of mannitol being a principal product of photosynthesis in the brown algae (Bidwell, 1958). Yamaguchi *et al.* (1966) have studied the mannitol content of *Eisenia bicyclis* both during photosynthesis and dark respiration. Photosynthesis in seawater containing $NaH^{14}CO_3$ resulted in very rapid incorporation of ^{14}C into the mannitol ($\sim 60\%$ of total in 5 min) while during subsequent incubation in the dark the specific activity of the hexitol was reduced by approximately 75% after 40 h (the mannitol content fell by about 50% in the same period). Thus mannitol is a very active metabolite in *E. bicyclis*.

A critical study of the utilization of mannitol by higher plants has been carried out by Trip and co-workers (1964) who fed ^{14}C-mannitol to 26 species from 17 families of higher plants and used evolution of $^{14}CO_2$ (in the dark) as an index of metabolism. It was found that 15 species from 11 different families respired mannitol to a certain extent, celery and four species of Oleaceae being among the most active. In the case of ash leaflets, metabolism of ^{14}C-mannitol was investigated further by following the appearance of other ^{14}C-labelled substances in the alcoholic extract: ^{14}C-fructose was the first to appear and was followed by sucrose, raffinose and a phosphorylated compound; no ^{14}C-mannose was detected. After 4 days, distribution of ^{14}C among components of ethanol-soluble fractions was the same whether ^{14}C-mannitol or ^{14}C-fructose was fed and, hence, these results indicate that the first step in mannitol catabolism is oxidation to fructose.

Mention has already been made of the accumulation of allitol in *Itea* leaves during photosynthesis. On respiration in the dark, however, the allitol content was observed to fall dramatically to less than half the initial value

TABLE IV

Patterns of labelling in isolated allitol and D-allulose after feeding (4 h) *Itea* leaves with ^{14}C-D-glucose and ^{14}C-D-fructose (Hough *et al.*, 1973)

	Location of ^{14}C (%)					
	Allitol			D-Allulose		
Substrate	C-1	C-2 → C-5	C-6	C-1	C-2 → C-5	C-6
D-[1-^{14}C]Glucose	45	24	28	47	19	30
D-[6-^{14}C]Glucose	28	17	53	28	16	53
D-[1-^{14}C]Fructose	59	18	22	56	16	27
D-[6-^{14}C]Fructose	27	12	64	28	13	53

after 23 h. From the yields of D-allulose (isolated as the di-*O*-isopropylidene derivative) it appears that the hexulose content of *Itea* leaves decreases during photosynthesis and increases during respiration in the dark; thus the inter-conversion, D-allulose ⇌ allitol, is indicated. This has been substantiated by studying the metabolism of: (a) ^{14}C-allitol and (b), specifically labelled D-glucose and D-fructose by *Itea* leaves (Hough *et al.*, 1973). The results for (a) showed that after respiration in the dark the specific activities of allitol and D-allulose were equal while the experiments with terminally labelled glucose or fructose yielded allitol and D-allulose with very similar patterns of labelling (Table IV).

Thus in the case of both mannitol and allitol, it appears that the first step in metabolism involves oxidation to a ketose and indeed it has usually been assumed that this is a general feature of polyol metabolism. This is also suggested by the frequent occurrence of polyol–ketose pairs in higher plants (Table V).

TABLE V

Simultaneous occurrence of alditol–ketose pairs in higher plants

Alditol	Ketose	Plant
D-Glucitol	D-Fructose	apple
Mannitol	D-Fructose	olive, ash
Allitol	D-Allulose	*Itea* spp.
D-*Glycero*-D-*manno*heptitol (volemitol)	D-*Altro*heptulose (sedoheptulose)	
D-*Glycero*-D-*galacto*heptitol (perseitol)	L-*Galacto*heptulose (perseulose)	avocado
D-*Glycero*-D-*gluco*heptitol (β-sedoheptitol)	Sedoheptulose	*Sedum* spp.
D-*Erythro*-D-*galacto*-octitol	D-*Glycero*-D-*manno*-octulose	avocado

V. Polyol Dehydrogenases and Metabolic Pathways

The conversion of an alditol to ketose will occur through the agency of a polyol dehydrogenase. To what extent have such enzymes been investigated? Most work has been done with fungi and cell-free enzyme systems catalysing interconversions of polyols and sugars (or their phosphates) (Lewis and Smith, 1967a). Some of the enzymes are NAD- and some NADP-dependent (as illustrated for penitol dehydrogenases (Fig. 1)) and it has been suggested that their concerted action can account for the interconversion of pentoses in

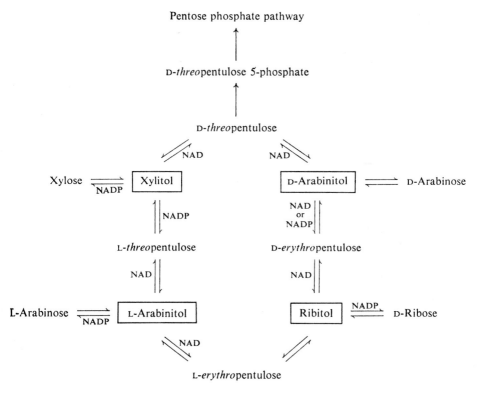

Fig. 1. Probable metabolic pathways in fungi.

fungi. The degree of specificity of the various polyol dehydrogenases is not clear but it has been pointed out by Touster and Shaw (1962) that earlier "rules" have proved far too simple.

Holligan and Jennings (1972) have recently shown how the relative importance of various routes (Fig. 2) for the synthesis of mannitol and arabinitol in *Dendryphiella salina* may be estimated from the specific activities of the polyols obtained after supplying D-[1-¹⁴C] and D-[6-¹⁴C] glucose to cultures of the fungus. There were two main facts involved in the argument;

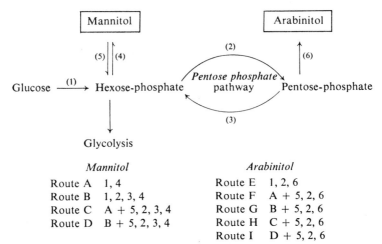

FIG. 2. Routes for the synthesis of mannitol and arabinitol from glucose by *Dendryphiella salina*.

(a), that oxidative reactions of the pentose phosphate pathway (reaction 2) lead to loss of ^{14}C from [1-^{14}C] hexose phosphate as $^{14}CO_2$ and (b), that if reversible formation of mannitol intervenes (reactions 4 and 5) there will be redistribution of label between C-1 and C-6 due to symmetry of the mannitol molecule (Fig. 3). Assuming substrate glucose has a specific activity of 1, specific activities of mannitol and arabinitol can be predicted (Table VI) (whenever intermediate synthesis and oxidation of mannitol occurs, the specific activity of C-6 is halved at the expense of C-1 and vice versa; this could also occur through resynthesis of triose phosphate during glycolysis but this was considered to have only a minor effect compared with the rapid synthesis and oxidation of mannitol).

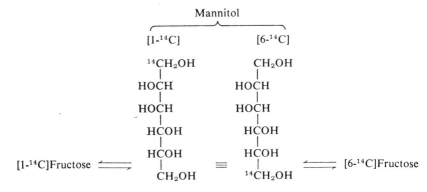

FIG. 3. Redistribution of label between C-1 and C-6 positions due to symmetry of mannitol.

TABLE VI

Predicted specific activity values of carbon atoms of mannitol and arabinitol derived from [1-^{14}C] and [6-^{14}C]glucose (Holligan and Jennings, 1972)

Route of synthesis[a]		D-[1-^{14}C]Glucose	D-[6-^{14}C]Glucose
Mannitol	A	1	1
	B	0	1
	C	$\frac{1}{2}$	$\frac{1}{2}$
	D	0	$\frac{1}{2}$
Arabinitol	E	0	1
	F	$\frac{1}{2}$	$\frac{1}{2}$
	G	0	$\frac{1}{2}$
	H	$\frac{1}{4}$	$\frac{1}{4}$
	I	0	$\frac{1}{4}$

[a] See Fig. 2.

The specific activity of the ^{14}C-mannitol was only slightly less than that of the substrate D-[6-^{14}C] glucose and hence routes C and D were considered to be of minor importance (and consequently, H and I were of minor importance in the synthesis of arabinitol). The main pathways were, therefore, considered to be A, B, E, F and G (Table VII). More than 60% of mannitol was formed by direct synthesis from glucose (route A) but the proportion formed via route B was greatly affected by the nitrogen source in the medium,

TABLE VII

Synthesis of mannitol and arabinitol in *Dendryphiella salina* from glucose in the presence of various nitrogen sources (Holligan and Jennings, 1972).

	G/NO$_3^-$ %	G/NH$_4^+$ %	G/glutamate %
Mannitol derived directly from substrate glucose (pathway A)[a]	61·7	72·4	61·2
Mannitol derived from substrate glucose via pentose phosphate pathway (pathway B)[a]	36·6	26·2	18·7
Mannitol derived from non-labelled substrate	1·7	1·4	20·1
Arabinitol derived directly from substrate glucose (pathway E)[a]	25·5	39·8	29·1
Arabinitol derived from substrate glucose via mannitol (pathways F and G)[a]	55·8	42·6	44·4
Arabinitol derived from non-labelled substrate	18·7	17·6	26·5

[a] See Fig. 2.

being highest in the presence of nitrate owing to the stimulation of the pentose phosphate pathway by this ion. A higher percentage of arabinitol was synthesized via mannitol when nitrate was in the medium and this was considered to reflect a more rapid turnover of the mannitol pool in the presence of nitrate.

There have been relatively few reports of isolation of polyol dehydrogenases from higher plants. Soluble dehydrogenases oxidizing mannitol and glucitol in the presence of NAD have been extracted from the berries of *Sorbus aucuparia* (Edson, 1953) while Kocourek *et al.* (1964) have obtained an acetone powder preparation of tobacco leaf which is highly specific for the NAD-dependent oxidation of L-arabinitol to L-*erythro*pentulose. It should be noted that there is no real evidence that pentose–pentulose interconversions in higher plants are mediated by pentitols as is the case in the fungi; in fact, isomerases which catalyse direct interconversions have been isolated from a variety of higher plant tissues (Pubols *et al.*, 1963).

With regard to conversion of aldoses to alditols in higher plants, Schradie (1966) has reported the isolation of an aldose reductase from the leaves of

TABLE VIII

Metabolism of D-[^{14}C]glucose by plum leaves (Anderson *et al.*, 1962)

Substrate	Location of ^{14}C in D-glucitol (%)		
	C-1	C-2 → C-5	C-6
D-[1-^{14}C]Glucose	70	9	21
D-[6-^{14}C]Glucose	16	6	78

Euonymus japonica, which is effective for the conversion of galactose to galactitol. Results of experiments with ^{14}C-glucose as substrate have suggested that the conversion of D-glucose to D-glucitol, occurs in the leaves of apple (Hutchinson *et al.*, 1959), sloe (Lewis, 1963) and plum (Anderson *et al.*, 1962). In the latter case, results of degradation experiments (Table VIII) clearly show that the glucitol is formed from specifically labelled glucose without appreciable redistribution of label.

It has been stressed that the first step in the catabolism of an alditol is usually considered to be oxidation to the corresponding ketose; however, some recent work concerning the metabolism of galactitol in leaves of *Euonymus japonica* indicates that this is not always the case. Galactose was shown to be the precursor of galactitol but the corresponding theoretical intermediate, tagatose (lyxohexulose), could not be detected (Schradie, 1966). When ^{14}C-galactitol was injected into midrib veins of *E. japonica* leaves, the specific activity of glucose formed after very short periods of time (15 sec–1·5 min) was consistently greater than that of the galactose,

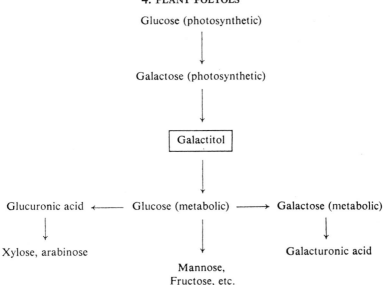

Glucose (photosynthetic)

Galactose (photosynthetic)

Galactitol

Glucuronic acid ⟵——— Glucose (metabolic) ———⟶ Galactose (metabolic)

Xylose, arabinose

Mannose,
Fructose, etc.

Galacturonic acid

FIG. 4. Proposed pathway for metabolism of galactitol in *Euonymus japonica*.

suggesting that glucose was the initial product of metabolism of galactitol (Bliss *et al*., 1972). A proposed metabolic scheme is presented in Fig. 4.

VI. ROLE OF ALDITOLS IN STORAGE OF REDUCING POWER AND
COENZYME REGULATION

In green plants reduced coenzymes are produced in photosynthesis. Lewis and Smith (1967a) have suggested that in plants lacking chlorophyll the frequent occurrence of the pentose-phosphate pathway may represent a substitute mechanism for the provision of reducing power which is stored as alditols; this could be accomplished, for example, by means of coupled systems such as:

glucose-6-P ⟶ ⟵ NADP ⟵ ⟶ alditol(-P)

gluconate-6P ⟵ ⟶ NADPH ⟶ ⟵ sugar(-P)

It may be noted that Strobel and Kosuge (1965) have observed that the level of mannitol in the mycelium in *Diplodia viticola* is closely correlated with that of glucose-6-phosphate dehydrogenase activity but *not* that of mannitol dehydrogenase activity.

A further way in which alditols may perform a regulatory function is through "transhydrogenase" activity (Touster and Shaw, 1962) when the synthesis of an alditol from an aldose is dependent on a different coenzyme

from its conversion to a ketose. A pair of polyol dehydrogenases could, therefore, serve as a regulatory system between NADPH and NAD as, for
example:

$$
\begin{array}{c}
\text{xylose} \quad \text{NADPH} \\
\text{NAD} \quad \text{xylitol} \quad \text{NADP} \\
\text{NADH} \quad \text{D-}threo\text{pentulose}
\end{array}
$$

VII. ALDITOLS IN TRANSLOCATION

Alditols appear to be the only carbohydrates, in addition to sucrose
and other non-reducing oligosaccharides, that are translocated in phloem
of higher plants. Thus, D-glucitol has been found in phloem exudates
of *Prunus serotina* (Rosaceae) and Webb and Burley (1962) report translocation of D-glucitol in apple. The most extensive studies on translocation
have been carried out by Trip *et al.* (1965) using lilac and white ash plants
and included the introduction of $^{14}CO_2$ into upper attached leaves during
photosynthesis, after which the plants were cut into five sections and separately
analysed. In these experiments more than ten carbohydrates became labelled
with ^{14}C in the leaves but in the stem and roots only verbascose, stachyose,
raffinose, sucrose and mannitol carried appreciable label. Since all the more
abundant non-reducing sugars of lilac and white ash but none of the reducing
sugars (with the possible exception of maltose) were translocated, it appears
that the non-reducing property of a carbohydrate is related to its function as a
transport material. Lewis and Smith (1967a) have suggested that since reducing
sugars are not translocated in the phloem, any diurnal conversion of an alditol
to its corresponding ketose (for example, allitol to D-allulose in *Itea* leaves)
may represent a control of carbohydrate movement during darkness.

VIII. CONCLUSION

We have seen that alditols are of widespread occurrence in diverse groups of
plants and this suggests they are of metabolic importance. In green plants they
appear to be direct products of photosynthesis and in this respect it may be
noted that the phosphates of the monosaccharides corresponding to erythritol,
ribitol, mannitol, glucitol, volemitol and β-sedoheptitol occur in the photosynthetic carbon cycle and may, therefore, be the respective alditol precursors.
The first step in alditol metabolism has usually been formulated as oxidation
to a ketose, although one apparent exception has been quoted.

Probable functions of plant alditols have been considered and include
storage of carbohydrate and reducing power, regulation of coenzymes and a
physiological role as agents of translocation.

Although polyol dehydrogenases have been isolated from a range of fungi our knowledge of enzymes concerned in alditol conversions in green plants is extremely limited, to say the least. Future research needs to be directed towards the isolation and characterization of these polyol dehydrogenases and will no doubt lead to a much better understanding of the metabolism and functions of plant alditols and their relationship with other storage compounds.

REFERENCES

Anderson, J. D., Andrews, P. and Hough, L. (1961). *Biochem. J.* **81**, 149.
Anderson, J. D., Andrews, P. and Hough, L. (1962). *Biochem. J.* **84**, 140.
Asai, T. (1937). *Jap. J. Bot.* **8**, 343.
Bednar, T. W. and Smith, D. C. (1966). *New Phytol.* **65**, 211.
Bidwell, R. G. S. (1958). *Can. J. Bot.* **36**, 337.
Black, W. A. P. (1950). *J. mar. biol. Ass. U.K.* **29**, 45.
Bliss, C. A., Hamon, N. W. and Lukaszewski, T. P. (1972). *Phytochemistry* **11**, 1695.
Boney, A. D. (1965). *Adv. mar. Biol.* **3**, 105.
Drew, E. A. (1966). D. Phil. thesis, University of Oxford.
Edson, N. L. (1953). *Report 29th meeting Aust. N.Z. Ass. Advance. Sci.*, Sydney, Australia, p. 281.
Holligan, P. M. and Jennings, D. H. (1972). *New Phytol.* **71**, 1119.
Hough, L. and Stacey, B. E. (1966a). *Phytochemistry* **5**, 215.
Hough, L. and Stacey, B. E. (1966b). *Phytochemistry* **5**, 171.
Hough, L., Shankar Iyer, P. N. and Stacey, B. E. (1973). *Phytochemistry* **12**, 573.
Hutchinson, A., Taper, C. D. and Towers, G. H. N. (1959). *Can. J. Biochem. Physiol.* **37**, 901.
Kocourek, J., Ticha, M. and Kŏstíř, J. V. (1964). *Archs Biochem. Biophys.* **108**, 349.
Lewis, D. H. (1963). D. Phil. thesis, University of Oxford.
Lewis, D. H. and Smith, D. C. (1967a). *New Phytol.* **66**, 143.
Lewis, D. H. and Smith, D. C. (1967b). *New Phytol.* **66**, 185.
Lindberg, B., Misiorny, A. and Wachmeister, C. A. (1953). *Acta chem. scand.* **7**, 591.
Lindberg, B. and Paju, J. (1954). *Acta chem. scand.* **8**, 817.
Nuccorini, R. (1930). *Ann. Chim. appl.* **20**, 535.
Plouvier, V. (1963). *In* "Chemical Plant Taxonomy" (ed. by T. Swain), p. 313. Academic Press, New York and London.
Pubols, M. H., Zahnley, J. C. and Axelrod, B. (1963). *Pl. Physiol.* **38**, 457.
Quillet, M. (1957). *Bull. Lab. marit. Dinard.* **43**, 119.
Sakai, A. (1961). *Nature, Lond.* **189**, 416.
Schradie, J. (1966). Ph.D. Thesis, University of Southern California, University Microfilms Inc., Ann Arbor.
Strobel, G. A. and Kosuge, T. (1965). *Archs Biochem. Biophys.* **109**, 622.
Touster, O. and Shaw, D. R. D. (1962). *Physiol. Rev.* **42**, 181.
Trip, P., Krotkov, G. and Nelson, C. D. (1963). *Can. J. Bot.* **41**, 1005.
Trip, P., Krotkov, G. and Nelson, C. D. (1964). *Am. J. Bot.* **51**, 828.
Trip, P., Nelson, C. D. and Krotkov, G. (1965). *Pl. Physiol.* **40**, 470.
Webb, K. L. and Burley, J. W. A. (1962). *Science, N.Y.* **137**, 766.
Yamaguchi, T., Ikawa, T. and Nisazawa, K. (1966). *Pl. Cell Physiol.* **7**, 217.

CHAPTER 5

Some Aspects of Sucrose Metabolism

*Department of Biochemistry, Royal Holloway College, University of London,
England*

I. Introduction 61
II. Sucrose Metabolism 62
    A. Sucrose Synthetase 64
    B. Sucrose Phosphate Synthetase 65
    C. Sucrose Phosphatase 65
    D. UDP-Glucose Pyrophosphorylase 66
III. Sites of Sucrose Metabolism in Sugar Cane Leaves 66
    A. Carbon Dioxide Fixation in HSK Plants 66
    B. Anatomy of the Sugar Cane Leaf 67
    C. Isolation of Sugar Cane Chloroplasts 67
    D. Methods of Separating Dimorphic Chloroplasts from Sugar Cane . 70
    E. Carbohydrate-Metabolizing Enzymes in Sugar Cane Chloroplasts . 72
Acknowledgement 80
References 80

I. INTRODUCTION

Sucrose is a non-reducing disaccharide of α-D-glucopyranose and β-D-fructofuranose joined by a $(1 \rightarrow 2)$-linkage.

sucrose

Sucrose is ubiquitous in green plants (Edelman, 1971) and is of great economic importance mainly as food material. Commercial sucrose is derived from plant material and is refined to a very pure state. Most of its world production comes from two sources: sugar cane, which supplies 55% of the world requirement for sucrose, and sugar beet which supplies 45% (Yudkin,

1971). However, very little is known about the pathways and the control mechanisms involved in the biosynthesis of sucrose in these plants (Walker 1971). It has been known for many years that green leaves can convert carbon dioxide to sucrose in the presence of light but the localization of the enzymes involved in the biosynthesis of the disaccharide from hexose phosphates has been a matter of conjecture. For a number of years many laboratory groups tried unsuccessfully to isolate chloroplasts capable of fixing CO_2 into sucrose. Gibbs *et al.* (1967) were the first to demonstrate this reaction using spinach chloroplasts but there are indications that seasonal and finely balanced control mechanisms are involved in this particular system. It is also generally accepted that the sucrose formed in the chloroplasts is the major translocatory product which is transported from the leaves via the phloem to the various plant organs where it can be utilized or stored (Spanner, 1971).

II. SUCROSE METABOLISM

The enzymes of the Calvin cycle can convert CO_2 into fructose 6-phosphate. This in turn is easily converted to glucose 6-phosphate and then glucose 1-phosphate by the action of the commonly occurring enzymes, glucose 6-phosphate isomerase and phosphoglucomutase, respectively.

There are two possible routes for the synthesis of sucrose from glucose 1-phosphate (Fig. 1). The first enzyme involved, UDP-glucose pyrophosphoryl-ase is common to both pathways. It catalyses the reaction:

$$\text{UTP} + \alpha\text{-D-glucose 1-phosphate} \rightleftharpoons \text{UDP-glucose} + \text{PP}_i$$

The following alternatives for the biosynthesis of sucrose from UDP-glucose have been suggested (Leloir and Cardini, 1955). The first involves the enzymes sucrose phosphate synthetase (UDP-D-glucose:D-fructose 6-phosphate 2-glucosyl transferase) and sucrose phosphatase (sucrose 6-phosphate phospho-hydrolase). The second route requires sucrose synthetase (UDP-D-glucose: D-fructose 2-glucosyl transferase). The sucrose phosphate synthetase pathway is now generally accepted as the major route for the biosynthesis of sucrose in plants. This belief is largely based on equilibrium data: the equilibrium constant for sucrose phosphate synthetase favours sucrose phosphate synthesis and, in addition, the phosphatase step in the pathway is virtually irreversible under normal physiological conditions. Therefore, sucrose phosphate is probably broken down as it is formed and the pathway can only operate in the direction of sucrose synthesis (Mendicino, 1960). However, the equilibrium constant for sucrose synthetase in the direction of sucrose synthesis is reported to be between 1 and 8, at pH 7·4, depending on the source of the enzyme and the conditions of the experiment (Leloir and Cardini, 1955; Cardini *et al.*, 1955). The reaction is thus freely reversible and it has been suggested that *in vivo* the enzyme is mainly responsible for the synthesis of nucleotide diphosphate glucose derivatives from sucrose, i.e. a reversal of sucrose synthesis.

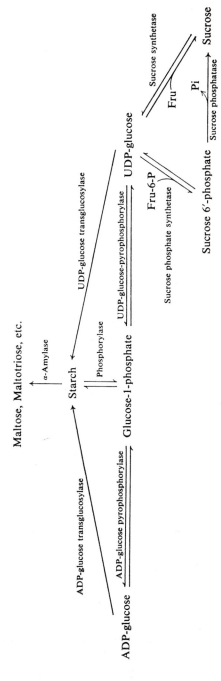

FIG. 1. General pathways for the metabolism of sucrose.

A. SUCROSE SYNTHETASE

This enzyme (UDP-D-glucose:D-fructose 2-glucosyl transferase) catalyses the following reaction:

$$\text{UDP-glucose} + \text{D-fructose} \rightleftharpoons \text{sucrose} + \text{UDP}$$

Grimes *et al.* (1970) have partially purified this enzyme from *Phaseolus aureus* (mung bean) seedlings and have shown that the enzyme is not specific for UDP-glucose but can utilize several other NDP-D-glucose derivatives for sucrose synthesis. The activities relative to UDP-glucose (100%) were shown to be: ADP-glucose, 27·7%; TDP-glucose, 10·2%; CDP-glucose, 3·1% and GDP-glucose, 2·6%. Sucrose synthetases from other sources appear to have similar properties in that they do not exhibit absolute specificity for UDP-glucose. It should be noted that UDP-glucose effectively inhibits the synthesis of sucrose by sucrose synthetase from other NDP-glucose derivatives. Delmer (1972a) has purified sucrose synthetase from *P. aureus* to homogeneity and has found that the enzyme has a high molecular weight (375 000 by ultra-centrifugation) and appears to consist of four identical subunits (mol wt 94 000). She has examined the kinetics of the enzyme and the equilibrium constant in the direction of UDP-glucose synthesis is 0·15 (pH 7·5, 25°). This agrees with previous data obtained with the enzyme when the reaction was measured in the direction of sucrose synthesis. When UDP is saturating the K_m for sucrose is 17 mM and when sucrose is saturating the K_m for UDP is 0·19 mM. The relative V_{max} values for the synthesis of the nucleoside diphosphate sugars agree closely with those found by Grimes *et al.* (1970) for the synthesis of sucrose (see above).

Delmer (1972b) has further investigated regulation of the enzyme using the homogeneous preparation from *P. aureus* and has found differences in properties when the enzyme is assayed in the synthetic and degradative directions. For example, NADP, iodoacetic acid, gibberellic acid and pyrophosphate all activate sucrose degradation but inhibit sucrose synthesis whereas fructose 1-phosphate and Mg^{2+} ions inhibit sucrose degradation. Pyrophosphate and fructose 1-phosphate are only effective when accompanied by Mg^{2+} ions. All these effectors (except gibberellic acid) were tested at a relatively high concentration (2 mM) so the physiological significance of the results can be questioned. Nevertheless, this work suggests that sucrose synthetase is subject to a number of complex regulatory factors. Other metabolites, notably glucose 1-phosphate, glucose 6-phosphate, a number of nucleotides and some intermediates in the Calvin cycle have no effect on the activity of the enzyme.

Murata (1971) has shown that the purified synthetase from sweet potato also exhibits different characteristics depending on whether sucrose synthesis or sucrose degradation is assayed. This high molecular weight enzyme is subject to allosteric regulation: the sucrose synthesis reaction shows the normal hyperbolic saturation curves whereas in the case of sucrose degradation sigmoidal saturation curves typical of allosteric enzymes are obtained. Thus, *in vivo*, the enzyme may only degrade sucrose when the concentration of the disaccharide is high (> 10 mM?).

B. SUCROSE PHOSPHATE SYNTHETASE

This enzyme (UDP-D-glucose:D-fructose 6-phosphate 2-glucosyl transferase) was first discovered in wheatgerm by Leloir and Cardini (1955). It catalyses the following reaction:

$$\text{UDP-glucose} + \text{fructose 6-phosphate} \rightleftharpoons \text{sucrose 6'-phosphate} + \text{UDP}$$

The enzyme appears to be specific for UDP-glucose and fructose 6-phosphate. Preiss and Greenberg (1969) have examined the kinetics of a purified enzyme from wheatgerm and found that the activity is regulated by the substrate concentration. The saturation curves for both fructose 6-phosphate and UDP-glucose are sigmoidal in the presence and absence of Mg^{2+}. $MgCl_2$ was found to stimulate the maximum velocity two-fold and decrease the apparent affinity for UDP-glucose three-fold. They suggest that the enzyme, and hence sucrose biosynthesis, is under substrate regulation *in vivo*.

Sucrose phosphate synthetase from *Vicia faba* cotyledons has been purified by Fekete (1971). This enzyme preparation contains an activating factor which is removed by freezing and thawing. The enzyme has very little activity in the absence of this factor but it is stimulated by citrate, some dicarboxylic acids and protamine. The saturation curves for these effectors are sigmoidal. In the presence of activator, citrate is inhibitory at low concentrations but reactivates the enzymes at high concentrations. When the enzyme is devoid of the factor, nucleotides and inorganic phosphate cause activation but with the activator–enzyme complex inhibition occurs with these compounds. The native and physiological significance of the naturally occurring activator is not known but it is clear that the enzyme is subject to complex metabolic regulation.

C. SUCROSE PHOSPHATASE

This enzyme (sucrose 6'-phosphate phosphohydrolase) catalyses the hydrolysis of sucrose 6'-phosphate to sucrose and inorganic phosphate. Hawker (1966) has detected sufficient sucrose phosphatase activity in sugar cane leaves to support the theory that sucrose phosphate is an intermediate in the biosynthesis of sucrose. In further work on the enzyme derived from stem and leaf tissues (Hawker and Hatch, 1966; Hawker, 1967) it was found that the enzyme requires Mg^{2+} but is inhibited by Ca^{2+}, Mn^{2+}, inorganic phosphate, pyrophosphate and sucrose. The inhibition of the enzyme by sucrose is partially competitive ($K_i = 10$ mM) and this may be of some significance in the control of sucrose synthesis in cane leaves.

D. UDP-GLUCOSE PYROPHOSPHORYLASE

The pyrophosphorylase (UTP:α-D-glucose 1-phosphate uridylyl transferase) catalyses the reaction:

$$\text{UTP} + \text{glucose 1-phosphate} \rightleftharpoons \text{UDP-glucose} + \text{PP}_i$$

The enzyme was first purified from *P. aureus* by Ginsburg (1958) who showed that it is specific for α-D-glucose 1-phosphate and UTP and that Mg^{2+} ions are required for activity. More recently, Gustafson and Gander (1972) obtained a purified preparation of the enzyme from *Sorghum vulgare* seedlings. They found that the saturation curve for UDP-glucose synthesis was sigmoidal but in the case of UDP-glucose breakdown it was hyperbolic.

III. SITES OF SUCROSE METABOLISM IN SUGAR CANE LEAVES

A. CARBON DIOXIDE FIXATION IN HSK PLANTS

Hatch and Slack (1966) demonstrated that the β-carboxylation reaction is of great importance in the fixation of CO_2 by sugar cane and other tropical plants such as *Zea mays*. The initial products of CO_2 fixation in sugar cane leaves are malate, oxaloacetate and aspartate whereas 3-phosphoglycerate is the product of photosynthesis in plants which fix CO_2 via the Calvin cycle.

The Australian group showed that in sugar cane the CO_2 is first fixed by phosphoenolpyruvate carboxylase which converts phosphoenolpyruvate to oxaloacetate. This latter keto acid is then reduced to malate, by a NADP dependent malate dehydrogenase, and converted to aspartate by aspartate aminotransferase. 3-Phosphoglycerate is formed at a later stage and this is followed by the production of hexose monophosphates, hexose diphosphates, sucrose and finally starch. It is believed that the initial products of CO_2 fixation by phosphoenol pyruvate carboxylase are transported to the site of the Calvin cycle enzymes where decarboxylation and refixation of the CO_2 by ribulose diphosphate carboxylase occurs (Hatch and Slack, 1970). Plants, such as sugar cane, which fix CO_2 by this method (HSK plants; Walker and Crofts, 1971) possess two structurally different types of chloroplast and are generally very resistant to photosaturation and lack detectable photorespiration: their carbon dioxide compensation points are below 5 ppm.

In our own studies (D. R. Davies and J. B. Pridham, unpublished results), which will now be described, we have examined the two types of chloroplasts in sugar cane leaves for enzymes concerned with sucrose and starch metabolism in an endeavour to determine whether these organelles play different roles.

B. ANATOMY OF THE SUGAR CANE LEAF

HSK plants have a distinctive leaf anatomy. A transverse section across a sugar cane leaf (Fig. 2) shows that the vascular bundles run parallel to the mid-vein of the leaf. The vascular bundle is surrounded by a single layer of cells called the bundle sheath and this, in turn, is surrounded by the dark green mesophyll cells. Some colourless bulliform cells are also visible.

A low power electron micrograph (Fig. 3) of a sugar cane leaf shows that the vascular bundle is surrounded by the inner bundle sheath and around this there is a single layer of cells, the bundle sheath. The bundle sheath chloroplasts of leaves that have been kept in the dark become packed with starch after illumination for 6 h. On the other hand only a few starch grains are visible in the mesophyll chloroplasts. The bundle sheath cells are surrounded by very thick cell walls whereas the mesophyll cell walls are comparatively thin. A cross section through adjacent bundle sheath and mesophyll

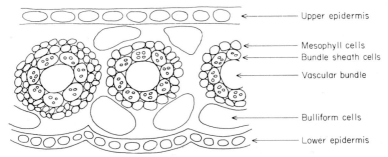

Fig. 2. Diagrammatic representation of a transverse section through a sugar cane leaf.

cells is shown in Fig. 4. The larger bundle sheath chloroplasts (5 μm diameter) have prominent starch grains whereas the mesophyll chloroplasts do not. The mesophyll chloroplasts appear to be typical plant chloroplasts bounded by a distinct outer membrane. They have prominent grana stacks linked by stroma lamellae. The spaces in between the membranes are filled with the soluble protein or stroma. The bundle sheath chloroplasts have single membranes or stroma lamellae, which appear to run parallel to each other. There have been no obvious grana stacks in any of the bundle sheath chloroplasts we have examined. The starch grains lie in the stroma between the stroma lamellae. One interesting feature is the existence of possible channels of translocation, the plasmodesmata, between the bundle sheath and the mesophyll cells. These observations regarding storage of starch confirm those of Laetsch et al. (1965, 1969).

C. ISOLATION OF SUGAR CANE CHLOROPLASTS

There are two major problems involved in the isolation of chloroplasts from sugar cane leaf. The first is the toughness of the tissue and the difficulty

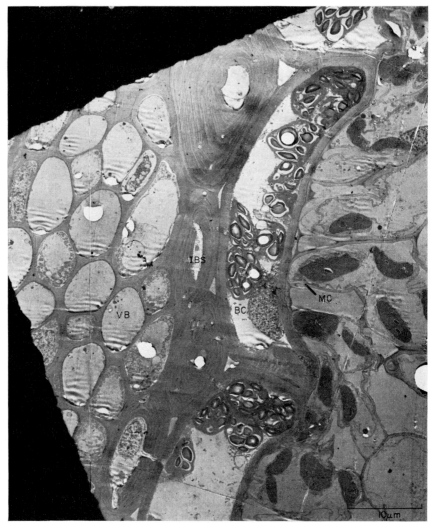

FIG. 3. Transverse section through a sugar cane leaf. IBS, inner bundle sheath cells; VB, vascular bundle; BC, bundle sheath cells; MC, mesophyll cells. (D. R. Davies and J. B. Pridham, unpublished results.)

of breaking down cell walls without rupturing the chloroplasts during the maceration procedure: the cell walls are especially thick around the bundle sheath cells (see Fig. 2). This problem can be partially overcome by maceration for short periods but this reduces the yield of bundle sheath chloroplasts.

The second major problem is the high levels of o-diphenols (mainly chlorogenic acid) and phenolases which are released during the isolation procedure and react to form enzyme inhibitors (Baldry *et al.*, 1970). In the

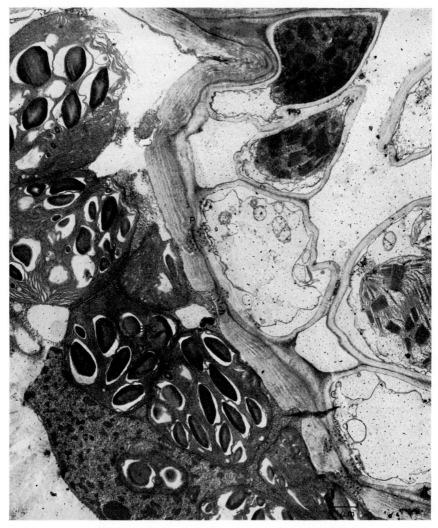

FIG. 4. Cross section through adjacent bundle sheath and mesophyll cells from sugar cane leaf. MC, mesophyll chloroplasts; BSC, bundle sheath chloroplasts; P, plasmodesmata. (D. R. Davies and J. B. Pridham, unpublished results.)

present investigation three possible methods were examined in an attempt to overcome this difficulty:

(a) removal of phenolic substrates by absorption with polyvinyl pyrolidone;
(b) removal of quinones by reduction with sodium isoascorbate;
(c) inhibition of phenolase by adding a thiol reagent such as cysteine or thioglycollate.

Generally, method (c) resulted in chloroplast preparations with the highest enzyme activities.

D. METHODS OF SEPARATING DIMORPHIC CHLOROPLASTS FROM SUGAR CANE

The main method used in the present study was the aqueous technique of Baldry *et al.* (1968). This method involves the grinding of the leaf tissue in isotonic sucrose, filtering the resulting brei through muslin and then harvesting the chloroplasts by rapid centrifugation. The resulting pellet is resuspended in isotonic sucrose solution, layered over 50% (w/w) aqueous sucrose solution and then centrifuged at 1000g for 10 min. The mesophyll chloroplasts accumulate at the interface of the two sucrose solutions whereas the starch-filled bundle sheath chloroplasts become pelleted at the bottom of the tube. The results of the protein and chlorophyll assays on chloroplast fractions obtained by this technique in a typical experiment are shown in Table I.

TABLE I

Protein and chlorophyll content of chloroplasts fractionated by the aqueous technique (D. R. Davies and J. B. Pridham, unpublished results)

Chlorophyll and protein content (μg per fraction)	Bundle sheath chloroplast fraction	Mesophyll chloroplast fraction
Chlorophyll	612	5400
Water soluble protein	514	392
NaOH soluble protein	1208	420
Total protein	1722	812

The low protein/chlorophyll ratio in the mesophyll chloroplast fraction indicates that a large proportion of these organelles were broken by the fractionation procedure. This was confirmed by examination of the fraction by electron microscopy which showed that the fraction consisted mainly of chloroplasts similar to those found in Calvin-type plants (see Fig. 5) but many of these organelles had lost their outer membranes and some stroma protein during the isolation procedure.

The bundle sheath chloroplast fraction (Fig. 6) obtained by the aqueous technique has more intact chloroplasts than the mesophyll fraction but there is some contamination of the former by cell wall material, ruptured mesophyll chloroplasts and individual starch grains. The latter are presumably derived from bundle sheath chloroplasts.

Enzyme preparations were made from both fractions by homogenizing the organelles in dilute buffer, to extract the soluble protein, and then centrifuging at 100 000g. The protein was precipitated with ammonium sulphate (80%) and the precipitate was resuspended in buffer and dialysed against buffer.

The second method used occasionally in the present study for separating

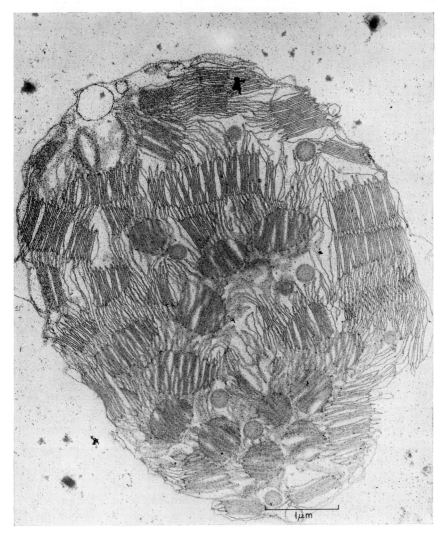

Fig. 5. Mesophyll chloroplast isolated by the aqueous technique. (D. R. Davies and J. B. Pridham, unpublished results.)

chloroplasts was the non-aqueous method of Slack (1969). This involves grinding lyophilized leaves in a carbon tetrachloride:hexane mixture (ρ = 1·25) and fractionating the chloroplasts by centrifugation in a non-aqueous medium of increasing density. On the whole the bundle sheath chloroplasts sediment at densities greater than 1·40 and the mesophyll chloroplasts at densities below 1·30. This technique conserves more stroma protein than the aqueous procedure. Soluble protein preparations were again made from these fractions as described above for the aqueous technique.

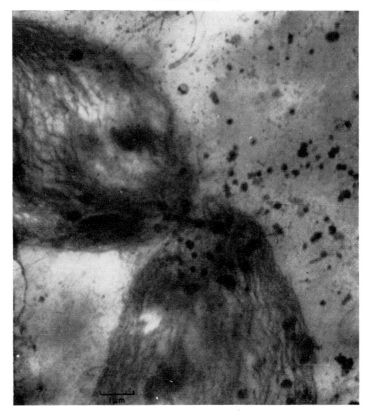

FIG. 6. Bundle sheath chloroplasts from sugar cane leaf isolated by the aqueous technique. (D. R. Davies and J. B. Pridham, unpublished results.)

E. CARBOHYDRATE-METABOLIZING ENZYMES IN SUGAR CANE CHLOROPLASTS

A number of enzymes of carbohydrate metabolism in cane mesophyll and bundle sheath chloroplasts were examined and the results are summarized in Table II (D. R. Davies and J. B. Pridham, unpublished results). Unless otherwise stated the organelles were fractionated by the aqueous method.

1. *Sucrose Synthetase*

This enzyme appeated to be very active in both types of chloroplast with a somewhat higher activity associated with the mesophyll fraction. This enzyme has previously been detected in sugar cane chloroplasts by Haq and Hassid (1965) who, because of the preparative method used, must have examined a fraction rich in mesophyll-type organelles. Bird *et al.* (1965) have also observed sucrose synthetase activity associated with chloroplasts obtained by the non-aqueous technique from tobacco. Huber and co-workers

TABLE II

Levels of enzyme activities[a] in sugar cane chloroplasts isolated by the aqueous technique (D. R. Davies and J. B. Pridham, unpublished results)

Enzyme	pH of assay	Bundle sheath chloroplast	Mesophyll chloroplast
Sucrose synthetase	7·5	4 810	5 930
Sucrose phosphate synthetase	7·5	1 660	1 350
UDP-glucose pyrophosphorylase	7·5	93·8	11·4
ADP-glucose pyrophosphorylase	7·5	9·3	1·0
Starch phosphorylase	6·5	26·7	1·0
UDP-glucose:starch transglucosylase (soluble)	8·4	< 0·1	< 0·1
UDP-glucose:starch transglucosylase (chloroplast bound)	8·4	< 0·1	< 0·1
ADP-glucose:starch transglucosylase (soluble)	8·4	16·2	61·8
ADP-glucose:starch transglucosylase (chloroplast bound)	8·4	12·9	15·9
Pyrophosphatase	8·3	108 000	206 000
Invertase (acid)	5·0	2·4	2·9
Invertase (neutral)	6·8	0·9	1·4
α-Amylase	6·8	12·5	5·6
UDP-glucose dehydrogenase	8·3	< 1.0	< 1·0
UDP-glucose 4′-epimerase	8·3	< 1·0	< 1·0
α-Galactosidase	5·0	< 1·0	< 1·0

[a] μmoles substrate converted per h per mg protein except in the case of α-amylase where activities are expressed in terms of μg apparent maltose released per h per μg protein.

(1969), however, could find little sucrose synthetase activity in either mesophyll or bundle sheath chloroplasts from *Zea mays*. This may be characteristic of the plant species or possibly due to enzyme instability. It should be noted that the enzyme from sugar cane is very labile and loses activity during isolation of the organelles unless cysteine (0·001 M) is added to the isolation medium. It is also interesting to find that Delmer and Albersheim (1970) detected synthetase activity in etiolated *Phaseolus aureus* plants but on exposure to light a rapid loss of synthetase activity occurred.

2. *Sucrose Phosphate Synthetase*

This enzyme was also found to be present in both bundle sheath and mesophyll chloroplasts from sugar cane. The specific activity of the enzyme isolated from the bundle sheath fraction was slightly higher than from the mesophyll fraction (see Table II). In this case the localization of the enzyme was also examined using non-aqueous chloroplast fractions (see p. 74). The distribution of enzyme in the various fractions again indicates that the enzyme is present in both bundle sheath and mesophyll chloroplasts (Fig. 7).

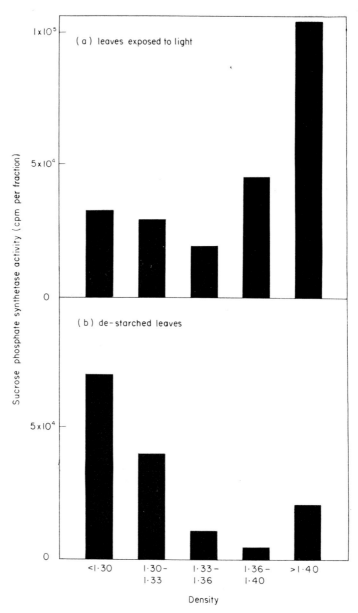

Fig. 7. Distribution of sucrose phosphate synthetase in chloroplast fractions from sugar cane obtained by the non-aqueous technique. (D. R. Davies and J. B. Pridham, unpublished results.)

Delmer and Albersheim (1970) have suggested that the enzyme in *Phaseolus aureus* chloroplasts is only active in the intact organelles. Washing the chloroplasts with 10% dimethyl sulphoxide was found to enhance the enzyme activity, presumably by increasing the permeability of the organelle to the substrates. In contrast, the enzyme from both types of sugar cane chloroplast can be extracted from the organelles with dilute buffer.

The theory of Delmer and Albersheim concerning the role of sucrose synthetase and sucrose phosphate synthetase is summarized in Fig. 8. They suggest that the former enzyme only occurs in non-photosynthetic tissues and here it is concerned with the synthesis of nucleotide diphosphate sugars and that sucrose phosphate synthetase is involved in the biosynthesis of sucrose in photosynthetic tissues. However, the theory does not explain the relatively high levels of sucrose synthetase which occur in sugar cane chloroplasts.

3. *UDP-glucose Pyrophosphorylase*

This enzyme was shown to be mainly localized in the bundle sheath chloroplast fraction of sugar cane but a small, but significant, level of the enzyme was also observed to be associated with the mesophyll chloroplasts (Table II). It was thought that this enzyme could be rate limiting in the synthesis of sucrose in chloroplasts and a study of the effects of various photosynthetic intermediates on the activity was, therefore, carried out. Fructose 6-phosphate, fructose 1,6-diphosphate, glucose 6-phosphate and 3-phosphoglycerate had no effect on activity at a level of 10^{-2} M. A gel filtration study on the enzyme showed that it was a small protein ($\sim 30\ 000$) and it appeared to be relatively stable when stored over long periods. There is, therefore, no evidence to suggest that the enzyme is allosteric and under the control of photosynthetic intermediates.

4. *ADP-glucose Pyrophosphorylase*

In contrast to the latter enzyme, this pyrophosphorylase does appear to be under allosteric control by photosynthetic intermediates in sugar cane. The enzyme is most active in the bundle sheath chloroplast fraction although a very low activity is also associated with the mesophyll fraction (Table II). The ratio of UDP-glucose pyrophosphorylase to ADP-glucose pyrophosphorylase in both fractions is approximately 10:1 but the activity of the latter enzyme is much more labile showing that the two activities probably result from two enzymic proteins.

The localization of the ADP-glucose pyrophosphorylase in organelle fractions prepared by the non-aqueous method (see p. 77) (Fig. 9) confirmed the results obtained by the aqueous separation procedure. Again most of the activity is associated with fractions of density $> 1\cdot40$ (i.e. bundle sheath chloroplasts) in contrast to sucrose phosphate synthetase (Fig. 7) which is associated with both mesophyll and bundle sheath chloroplasts.

D. R. DAVIES

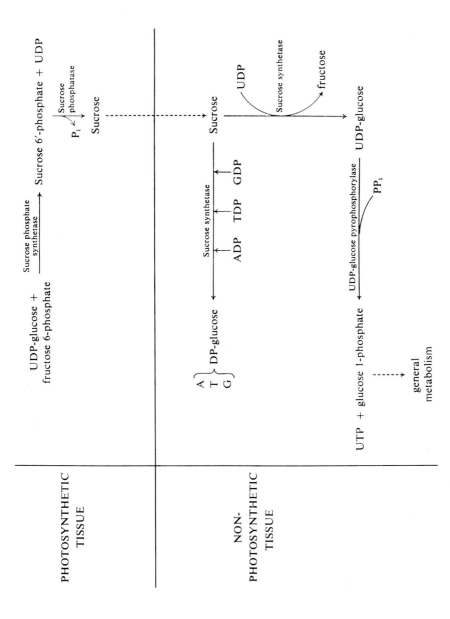

FIG. 8. The possible role of sucrose synthetase and sucrose phosphate synthetase in sucrose metabolism in *P. aureus* (cf. Delmer and Albersheim, 1970).

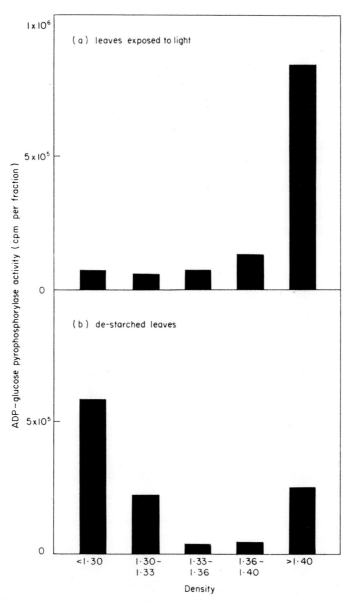

FIG. 9. Distribution of ADP-glucose pyrophosphorylase in chloroplast fractions from sugar cane obtained by the non-aqueous technique. (D. R. Davies and J. B. Pridham, unpublished results.)

Ghosh and Preiss (1966) and Sanwal *et al.* (1968) have examined ADP-glucose pyrophosphorylase in some detail and they have found that the purified enzyme from spinach and other plants shows allosteric properties. Generally, the enzyme derived from leaf tissue is activated by intermediates of the Calvin cycle such as 3-phosphoglycerate, fructose 6-phosphate and fructose 1,6-diphosphate. This activation is reversed by inorganic phosphate so that during photosynthesis when inorganic phosphate is being utilized in photophosphorylation and the levels of photosynthetic intermediates are increased, ADP-glucose synthesis is favoured and hence increased starch synthesis follows.

However, in sugar cane the first products of photosynthesis are malate, oxalate and aspartate. It was, therefore, of interest to find out if the enzyme from sugar cane is controlled by the same photosynthetic intermediates as the enzyme in plants which fix CO_2 via the Calvin cycle. The results of such a study using an enzyme preparation derived from a non-aqueous bundle sheath fraction from sugar cane is summarized in Table III which shows

TABLE III

The effect of photosynthetic intermediates and magnesium ions on pyrophos-phorylase activity

	Specific activity (nmol product/h/mg protein)	
	ADP-glucose pyrophosphorylase	UDP-glucose pyrophosphorylase
No additions	180	4480
+ 3-phosphoglycerate (3·3 mM)	701	4520
+ fructose 6-phosphate (3·3 mM)	788	4390
− Mg^{2+}	0	0

that ADP-glucose pyrophosphorylase, but not the corresponding UDP-glucose enzyme, is strongly activated by 3-phosphoglycerate and fructose 6-phosphate. The activation of ADP-glucose pyrophosphorylase by 3-phosphoglycerate was examined in more detail (Fig. 10) and was shown to be reversed by inorganic phosphate. Therefore, ADP-glucose pyrophosphorylase in sugar cane, an HSK plant, can probably be regulated by the same initial products of photosynthesis which are produced by Calvin-type plants.

It is interesting to note (Table III) that both pyrophosphorylases are completely inactive in the absence of Mg^{2+} ions. This may be of importance because there is a light-induced pumping of Mg^{2+} ions from the thylakoids into the stroma during photosynthesis (Dilley and Vernon, 1965). It has been estimated that the concentration of Mg^{2+} in the stroma changes from 10 mM

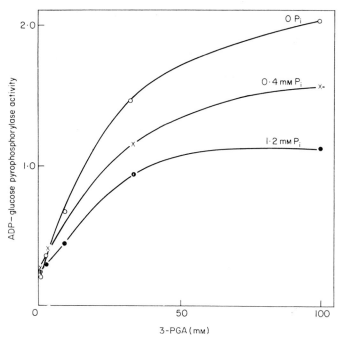

FIG. 10. Activation of ADP-glucose pyrophosphorylase by 3-phosphoglycerate in the presence of varying amounts of inorganic phosphate. (D. R. Davies and J. B. Pridham, unpublished results.)

in the light to a very low level in the dark (see Chapter 2). Thus the pyrophosphorylases may be controlled by changes in the Mg^{2+} concentration in the chloroplast stroma.

5. *Invertase*

This enzyme is not generally thought to be associated with chloroplasts (Bird *et al.*, 1965) and the present study has shown that there is very little "neutral" or "acid" activity associated with either bundle sheath or mesophyll chloroplasts from sugar cane (Table II). It is a cytoplasmic enzyme in this plant and the low activity associated with the chloroplast is probably derived from cytoplasmic contamination.

6. *Pyrophosphatase*

There is a high level of the enzyme associated with both types of cane chloroplast particularly in the case of the mesophyll fraction (Table II). This enzyme may be of importance in assisting the pyrophosphorylase-catalysed reactions in the direction of nucleoside diphosphate sugar synthesis and hence in the direction of both sucrose and starch synthesis.

7. *Starch Phosphorylase*

Bundle sheath chloroplasts appeared to be the main sites for this activity in cane leaves (Table II). The observations agree with the data published by Huber *et al.* (1969) on the distribution of the enzyme in *Zea mays*.

8. *ADP-glucose: Starch Transglucosylase*

Both bundle sheath and mesophyll chloroplasts of sugar cane possessed this activity (Table II). There appears to be more of the soluble form of the enzyme associated with the mesophyll fraction than with the bundle sheath organelles but the specific activities of the bound enzymes are similar in both fractions. No UDP-glucose transglucosylase activity could be detected in either cane chloroplast fraction (Table II).

This investigation of the location and regulatory properties of enzymes of carbohydrate metabolism in sugar cane leaves leads us to the conclusion that both types of chloroplast possess the necessary enzymes to synthesize sucrose from glucose 1-phosphate and also to degrade sucrose to hexose phosphate. Most of the starch is found in bundle sheath chloroplasts and it is in these organelles that the enzymes of starch metabolism, both synthetic and degradative, are localized. The starch that does occur in the mesophyll chloroplasts could arise from ADP-glucose via sucrose synthetase.

ACKNOWLEDGEMENT

We are greatly indebted to Tate and Lyle Ltd., for financial support and much helpful advice in connection with this study.

REFERENCES

Baldry, C. W., Bucke, C., Coombs, J. and Gross, D. (1970). *Planta* **94**, 107.
Baldry, C. W., Coombs, J. and Gross, D. (1968). *Z. Pflanzenphysiol.* **60**, 78.
Bird, I. F., Porter, H. K. and Stocking, C. R. (1965). *Biochim. biophys. Acta* **100**, 366.
Cardini, C. E., Leloir, L. F. and Chiriboga, J. (1955). *J. biol. Chem.* **214**, 149.
Delmer, D. P. (1972a). *J. biol. Chem.* **247**, 3822.
Delmer, D. P. (1972b). *Pl. Physiol.* **50**, 469.
Delmer, D. P. and Albersheim, P. (1970). *Pl. Physiol.* **45**, 782.
Dilley, R. A. and Vernon, L. P. (1965). *Archs Biochem. Biophys.* **111**, 365.
Edelman, J. (1971). *In* "Sugar" (J. Yudkin, J. Edelman and L. Hough, eds), p. 95. Butterworths, London.
Fekete, M. A. R. de (1971). *Eur. J. Biochem.* **19**, 73.
Ghosh, H. P. and Preiss, J. (1966). *J. biol. Chem.* **241**, 4491.
Gibbs, M., Latzko, E., Everson, R. G. and Cockburn, W. (1967). *In* "Harvesting the Sun" (A. San Pietro, F. Greer and T. J. Army, eds), p. 111. Academic Press, New York and London.
Ginsburg, V. (1958). *J. biol. Chem.* **232**, 55.

Grimes, W. J., Jones, B. L. and Albersheim, P. (1970). *J. biol. Chem.* **245**, 188.

Gustafson, G. L. and Gander, J. E. (1972). *J. biol. Chem.* **247**, 1387.

Haq, S. and Hassid, W. Z. (1965). *Pl. Physiol.* **40**, 591.

Hatch, M. D. and Slack, C. R. (1966). *Biochem. J.* **101**, 103.

Hatch, M. D. and Slack, C. R. (1970). *In* "Progress in Phytochemistry" (L. Rheinold and Y. Liwschits, eds), Vol. 2, p. 35. Interscience.

Hawker, J. S. (1966). *Phytochemistry* **5**, 1191.

Hawker, J. S. (1967). *Biochem. J.* **105**, 943.

Hawker, J. S. and Hatch, M. D. (1966). *Biochem. J.* **99**, 102.

Huber, W., de Fekete, M. A. R. and Ziegler, H. (1969). *Planta* **87**, 360.

Laetsch, W. M., Stetler, D. A. and Vlitos, A. J. (1965). *Z. Pflanzenphysiol.* **54**, 472.

Laetsch, W. M. and Price, I. (1969). *Am. J. Bot.* **56**, 77.

Leloir, L. F. and Cardini, C. E. (1955). *J. biol. Chem.* **214**, 157.

Mendicino, J. (1960). *J. biol. Chem.* **235**, 3347.

Murata, T. (1971). *Agric. biol. Chem.* **35**, 1441.

Preiss, J. and Greenberg, E. (1969). *Biochem. biophys. Res. Commun.* **36**, 285.

Sanwal, G. G., Greenberg, E., Hardie, J., Cameron, E. C. and Preiss, J. (1968). *Pl. Physiol.* **43**, 417.

Slack, C. R. (1969). *Phytochemistry* **8**, 1387.

Spanner, D. C. (1971). *In* "Sugar" (J. Yudkin, J. Edelman and L. Hough, eds), p. 110. Butterworths, London.

Walker, D. A. (1971). *In* "Sugar" (J. Yudkin, J. Edelman and L. Hough, eds), p. 103. Butterworths, London.

Walker, D. A. and Crofts, A. R. (1971). *A. Rev. Biochem.* **39**, 389 (1970).

Yudkin, J. (1971). *In* "Sugar" (J. Yudkin, J. Edelman and L. Hough, eds), p. 11. Butterworths, London.

CHAPTER 6

The Nature and Function of Higher Plant α-Galactosidases

J. B. PRIDHAM and P. M. DEY

Department of Biochemistry, Royal Holloway College,
University of London, England

I. INTRODUCTION

The glycosidases, enzymes which catalyse the hydrolysis and formation of glycosidic bonds, have been known for more than 150 years although in recent times there has been much more emphasis on the involvement of inorganic phosphate and nucleotide derivatives in the metabolism of glycosides.

Early workers who studied glycosidases were content to catalogue the various activities in plant cells and it soon became apparent that the situation was complex. The enzymes could easily be obtained as crude but highly active preparations and many laboratories in the period from the end of the Second World War to the 1960s were interested in their catalytic activities *in vitro* and, in particular, specificities, mechanisms of reaction and products resulting from transfer of glycosyl residues to acceptors other than water. Arguments also raged as to whether two types of glycosidase activity in a cell were associated with one or two protein molecules. Although it was generally believed that glycosidases played some role in carbohydrate metabolism in plants, particularly during germination, few detailed studies were ever attempted in this area.

In recent years there has been a renewed interest in glycosidases particularly in relation to use as specific hydrolytic agents, mechanisms of reaction and existence of multiple forms.

In our laboratory we are attempting to deduce the physiological role of glycosidases and an extensive examination of α-galactosidases in seeds, particularly from *Vicia faba*, is under way.

In this review an attempt has been made to summarize the preliminary results of this study. Readers requiring a more general review of α-galactosidases should refer to Dey and Pridham (1972).

II. STRUCTURE AND PROPERTIES OF α-GALACTOSIDASES

α-Galactosidases (α-D-galactoside galactohydrolases) were first reported in 1895 (Bau, 1895; Fischer and Lindner, 1895) and to date preparations of the enzyme have been obtained from approximately 60 different sources, mostly from seeds and fungi. α-Galactosidases catalyse the following reversible reaction:

(With most enzymes R may represent alkyl, aryl, monoglycosyl or polyglycosyl groups.)

Most forms of the enzyme are not absolutely specific for the glycon residue and will hydrolyse the structurally related α-D-fucopyranosides and β-L-arabinopyranosides (e.g. Dey and Pridham, 1969b). Water can be replaced by a number of organic galactose acceptors and in this way simple galactosides and oligosaccharides have been synthesized. The equilibrium for a α-galactosidase-catalysed reaction, as with other glycosidases, normally favours hydrolysis. In the case of an enzyme isolated from *Phaseolus vulgaris* by Tanner and Kandler (1968), however, transgalactosylation from galactinol to raffinose (with the formation of stachyose) occurs more readily than the hydrolysis of these galactosides. *De novo* synthesis of oligosaccharides is also possible when high concentrations of galactose are incubated with the enzyme (see Dey and Pridham, 1972).

As regards the chemical and physical nature of the enzymic proteins there have been relatively few studies owing to the paucity of highly purified α-galactosidase preparations. Only in one instance has the enzyme been crystallized (from the fungus *Mortierella vinacea*; Suzuki *et al.*, 1970) although apparently homogeneous preparations have been obtained from *Aspergillus niger* (Lee and Wacek, 1970), *Vicia faba* seeds (Dey and Pridham, 1969a; Dey *et al.*, 1971) and *Vicia sativa* seeds (Petek *et al.*, 1969). Some data on the amino acid compositions have been published in the case of enzymes from *M. vinacea* (a glycoprotein) and *V. sativa* and preliminary analyses of α-galactosidases from *V. faba* are discussed on p. 86 of this chapter.

The occurrence of multimolecular forms of the enzyme was first reported from Courtois' laboratory (Petek and Dong, 1961; Courtois and Petek, 1966; Courtois *et al.*, 1963). Extracts of seeds from *Coffea* sp. and *Plantago ovata* were both resolved into two active fractions by chromatography on alumina. Dey and Pridham (1968, 1969a) later showed that extracts of dormant *V. faba* seeds contained two α-galactosidases, differing in their molecular weights, and that the mature seeds from a number of other plant species also possessed two forms of the enzyme (Barham *et al.*, 1971). Multiple forms have also been detected in fungi by Lee and Wacek (1970) and Suzuki *et al.* (1970).

III. STUDIES WITH *Vicia faba* α-GALACTOSIDASES

A. MULTIMOLECULAR FORMS IN MATURE SEEDS

Our particular interest in α-galactosidases is a logical extension of early studies in Edinburgh (Pridham, 1958) on the occurrence of galactosylsucrose derivatives in *Vicia faba* (broad bean) seeds and seedlings.

In preliminary work a seven-stage procedure was used to obtain two molecular forms (I and II) of α-galactosidase which occur in mature bean seeds. They were purified 3660- and 337-fold, respectively, and both behaved as homogeneous protein preparations when examined by polyacrylamide gel electrophoresis. The molecular weights of I and II, estimated from Sephadex gel filtration studies, were 209 000 and 38 000, respectively, and enzyme I could be dissociated into six inactive protein fractions of varying molecular weights when treated with 6M-urea (Dey and Pridham, 1969a). The detailed kinetic properties of the two α-galactosidases were examined and information concerning the nature of the active site of I also obtained. Of particular interest was the specific activity of I which was significantly higher than that of II when several synthetic and naturally occurring α-D-galactosides were used as substrates (Dey and Pridham, 1969b). Enzyme I, but not II, was also observed to hydrolyse galactomannans which are found as reserve polysaccharides in some seeds (e.g. Table I; P. M. Dey, A. Khaleque and J. B. Pridham, unpublished results).

TABLE I

Hydrolysis of galactomannan (locust bean gum) by *Vicia faba* α-galactosidases
(P. M. Dey, A. Khaleque and J. B. Pridham, unpublished results)

α-Galactosidase	Galactomannan composition	
	% Galactose	% Mannose
Control	23	77
+ Enzyme I[a]	17	83
+ Enzyme II[a]	23	77

[a] 45 min incubations at 25°C.

In later work the two forms of the enzyme, obtained from Sephadex columns, were re-examined by ion exchange chromatography (Dey et al., 1971). No further resolution of either enzyme was affected with DEAE-cellulose columns. However, when enzyme II was applied to a column of CM-cellulose which was eluted stepwise with buffer of increasing pH, two active fractions (II^1 and II^2 eluting at pH 6·0 and pH 6·5, respectively) were obtained together with four inactive fractions. The amount of protein associated with the isoenzymes was approximately equal but the total activity of the II^1 peak was three times higher than that of the II^2 fractions. The properties of II^1 and II^2 are given in Table II. Elution of enzyme I from a CM-cellulose column gave rise to two protein peaks but only one, which was eluted at pH 4·5, possessed α-galactosidase activity.

Originally, α-galactosidase I from mature bean seeds was thought to be a glycoprotein containing 25% carbohydrate (Dey and Pridham, 1969a); however, a recent improved purification of the enzyme (P. M. Dey, A. Khaleque and J. B. Pridham, unpublished results) has resulted in a product

TABLE II

Kinetic properties of *Vicia faba* α-galactosidases II^1 and II^2 from mature seeds (P. M. Dey, A. Khaleque and J. B. Pridham, unpublished results)

Property	II^1	II^2
pH optimum[a]	3·0; 5·2	2·0; 5·5
V_{max} (μmol hydrolysed/min/mg protein)[a]	1·61	0·35
K_m (mM)[a]	0·97	0·33

[a] *p*-nitrophenyl α-D-galactoside substrate.

containing only 8·0% carbohydrate and a similar value was also obtained for enzyme II. Enzymes II^1 and II^2 have both been shown to contain approximately 4% carbohydrate. In all cases it will be necessary to demonstrate covalent bonding between carbohydrate and protein before the glycoprotein nature of the enzymes can be confirmed. In view of the high polysaccharide content of the initial seed extracts it is possible that the enzymes are physically contaminated with these polymers.

Preliminary amino acid analyses have been carried out on preparations of I, II^1 and II^2, purified by an eight-stage procedure which included Sephadex and CM-cellulose chromatography in the final stages. In the case of the preparation of enzyme II which was analysed, the final stage consisted of re-cycling through Sephadex. All the fractions were shown to be homogeneous when examined with an analytical ultracentrifuge. The recovery of amino acids (see Table III) was variable but the obvious similarity in the compositions of the four enzymes is undoubtedly of some significance.

Analysis of the data obtained from ultracentrifugation of all the above fractions gave values for their molecular weights and these are shown in

TABLE III

Amino acid compositions of α-galactosidases from *Vicia faba* seeds (P. M. Dey, A. Khaleque and J. B. Pridham, unpublished results)

Amino acid	Enzyme I Mature seeds (recovery 71·5%)	Enzyme II Mature seeds (recovery 90·1%)	Enzyme II[1] Mature seeds (recovery 81·7%)	Enzyme II[2] Mature seeds (recovery 90·3%)	Enzyme I Converted from Enzyme II in vitro (recovery 90·3%)	Enzyme II Germinated seed (recovery 92·5%)
			μmol/100 mg protein			
Lysine	14·5	15·5	14·7	15·1	15·6	15·2
Histidine	3·9	6·9	3·8	5·6	4·6	5·9
Arginine	10·8	12·0	10·3	10·3	7·7	10·0
Aspartic acid	34·8	32·4	30·0	32·6	32·4	33·0
Threonine	15·7	18·0	13·0	16·0	18·3	17·0
Serine	26·1	26·7	23·0	25·1	26·9	26·4
Glutamic acid	22·9	23·8	21·8	24·2	24·2	24·6
Proline	12·7	16·6	19·0	20·6	20·5	20·6
Glycine	25·5	26·0	24·1	25·1	30·3	25·4
Alanine	18·8	18·8	14·7	18·4	22·0	18·3
Cysteine	8·4	10·4	15·1	9·9	10·2	13·3
Valine	14·5	16·8	14·0	15·0	16·4	15·8
Methionine	1·1	1·7	1·7	2·7	1·0	3·0
Isoleucine	9·6	11·2	10·4	11·4	10·7	13·0
Leucine	19·9	19·6	16·8	19·6	24·1	21·8
Tyrosine	7·4	8·0	8·7	11·4	9·5	10·7
Phenyl alanine	9·6	11·0	11·0	12·3	10·6	13·5

Table IV together with those obtained by Sephadex gel-filtration. Both methods give similar molecular weights for enzymes II^1 and II^2, i.e. of the order of 40 000. However, although the gel-filtration method indicates that II^1, II^2 and II are of a similar molecular size this is not borne out by the sedimentation equilibrium technique which suggests that II has a higher molecular weight than either of its components. Similarly, the values obtained for the molecular weight of enzyme I by the two methods are quite different. These anomalies are, so far, unexplained. The carbohydrate contents of the protein fractions may have some bearing on the problem and, in addition, ultracentrifuge data suggest that an association phenomenon may exist in the case of enzyme I. The latter finds support in the observation that II can be converted to an enzyme closely resembling I *in vitro* (see p. 91).

TABLE IV

Analytical data for the four molecular forms of α-galactosidase from mature *Vicia faba* seeds (from P. M. Dey, A. Khaleque and J. B. Pridham, unpublished results)

| Enzyme | Molecular weight | | Carbohydrate content % (as glucose) | Potassium content (meq/mg protein) |
	Sedimentation equilibrium method	Sephadex gel filtration method		
I	160 400 ± 2850	209 000	8·0	40·0
II	54 340 ± 5225	38 000	7·8	10·0
II^1	45 730 ± 3073	38 000	4·2	3·0
II^2	43 390 ± 1409	38 000	4·3	0·5

The analytical data presented, so far, suggest that α-galactosidase II may be composed of an equimolar mixture of two active proteins, II^1 and II^2, of similar size together with other (four?) enzymically inactive components. As the carbohydrate contents of II^1 and II^2 are the same and the value is approximately 50% of that of enzyme II this may mean that the inactive protein fractions have a considerable amount of carbohydrate associated with them. The amino acid analyses also suggest that there is a close relationship between II and I (see also p. 91) and our working hypothesis is that monomers of II associate to form I and that *in vitro* there may be differing degrees of association.

In early studies on the purification of the multimolecular forms of α-galactosidase from *Vicia faba* seeds it was observed that passage of a crude mixture of I and II through a Sephadex column often resulted in a doubling of the total enzyme activity (Dey *et al.*, 1971). This was eventually traced to an activation of both forms by K^+ ion which was used in the column eluent. Maximum activation of an enzyme mixture prepared in the absence of K^+ was effected by 0·12 M KCl and the process was shown to be time dependent, the maximum increase in activity occurring only after preincubation of the enzyme preparation with KCl for several hours. K^+ could not be replaced by Na^+ although some activation was observed with NH_4^+ ions. After passing the enzyme mixture (prepared in the absence of exogenous K^+) through Sephadex G-100 the resulting fractions of α-galactosidase I and II could not be activated by K^+.

The endogenous bound K^+ in the enzymes prepared in the absence of the alkali metal ion was examined. Forms II^1 amd II^2 possessed a very low K^+ content and the level in enzyme I was four times that in enzyme II (Table IV).

B. CHANGES IN α-GALACTOSIDASES DURING SEED MATURATION

Examination of the Sephadex patterns and total activity of α-galactosidases in immature seeds has shown that the small, green beans contain only a low level of enzyme II and that this increases with maturation. Following this

TABLE V

α-Galactosidase activity in maturing *Vicia faba* seeds (P. M. Dey, P. R. Palan and J. B. Pridham, unpublished results)

Plant age	Average seed length (cm)	Enzyme activity (mUnit[a]/ml)	Protein (mg/ml)	Specific activity $\left(\dfrac{\text{mUnit/ml}}{\text{mg protein/ml}}\right)$
12	0·7	Nil	—	—
13(a)[b]	1·9	6·9	0·7	9·9
13½	2·2	9·5	0·9	0·6
17(b)[b]	2·9	28·8	2·3	12·5

Seeds were then removed from pods and stored at room temperature in a Petri dish:

Storage time (days)	Enzyme activity (mUnit/ml)	Protein (mg/ml)	Specific activity $\left(\dfrac{\text{mUnit/ml}}{\text{mg protein/ml}}\right)$
4	44·0	1·7	25·9
5(c)[b]	44·0	1·6	27·5
7	47·0	1·6	29·4

[a] One unit of activity is defined as the number of μmol of p-nitrophenyl α-D-galactoside hydrolysed per min at 30°C, pH 5·5.

[b] (a), (b) and (c) refer to Sephadex gel-filtration patterns (see Fig. 1).

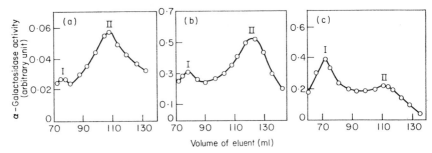

FIG. 1. Sephadex G-100 gel-filtration patterns of α-galactosidase activity in maturing *Vicia faba* seeds. The stages of maturation, (a), (b) and (c), are described in Table V (from Dey *et al.*, 1973).

stage a marked increase in the specific activity occurs (Table V) which can be related to the appearance of enzyme I and a decrease in the activity of enzyme II (Dey *et al.*, 1973; P. M. Dey, A. Khaleque and J. B. Pridham, unpublished results; see Fig. 1). At the present time it is not certain whether I is derived from II *in vivo*; the appearance of I and disappearance of II during

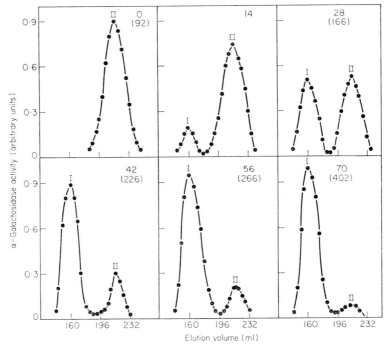

FIG. 2. Conversion of α-galactosidase II to a high molecular form (I?) *in vitro* at pH 5·5 (4°C). The solution was examined at intervals (figures at the top of each pattern represent time in days) on a Sephadex G-100 column and at the same time the specific activity (figures in parenthesis (mUnits/mg protein)) was measured (cf. Dey *et al.*, 1971).

maturation could be explained in a number of ways including the *de novo* synthesis of I coupled with the proteolysis of II. However, the analytical data suggest that there is a structural relationship between the two forms of α-galactosidases and investigations have shown that, *in vitro*, association of units of II may result in I or an enzyme closely resembling I.

A batch of broad bean seeds (Bunyard Exhibition Longpod) which appeared to be mature were observed to be abnormal in that they contained only enzyme II. These particular seeds were used as a source of partially purified enzyme II which could be prepared by simple precipitation procedures without the use of gel filtration. When such a preparation was stored at

TABLE VI

Comparison of α-galactosidase I from mature *Vicia faba* seeds and enzyme formed from α-galactosidase II *in vitro* (P. M. Dey, A. Khaleque and J. B. Pridham, unpublished results)

Properties	α-Galactosidase I (mature seeds)	α-Galactosidase (from α-galactosidase II)
Specific activity (milliunits/mg protein; substrate *p*-nitrophenyl α-D-galactoside)	65 631	53 240
Elution volume (ml) (Sephadex G-200 column; 86 × 2·5 cm)	190	190
Elution pH (CM-cellulose chromatography)	4·5	4·5
Molecular weight		
Sephadex Gel-filtration method	209 000	209 000
Sedimentation equilibrium method	160 040 ± 2850	118 000 ± 3844
Carbohydrate content (as μg glucose/mg protein)	80	58
E 280/260	1·34	1·56
E 1% 280	18	21
Potassium content (meq/mg protein)	40	35
pH optimum	2·0; 5·5	2·0; 5·5
K_m (mM; substrate, *p*-nitrophenyl α-D-galactoside)		
pH 2·0	0·44	0·40
pH 5·5	0·44	0·39
V_{max} (μmol/min/mg protein; substrate, *p*-nitrophenyl α-D-galactoside)		
pH 2·0	14·62	15·65
pH 5·5	24·15	27·11
Action on locust bean (see Table I) and Tara galactomannans	Galactose released	Galactose released

4° (pH 5·5) over a period of several days it was noted that the specific activity of the solution slowly increased and that this was accompanied by the formation of a high molecular weight α-galactosidase and the loss of II (Fig. 2). Further study of this phenomenon revealed that the conversion could be accelerated by raising the pH to 7·0 and the temperature to 25°. Conversion to the high molecular weight form was prevented by passage of enzyme II through a Sephadex G-100 column prior to incubation.

A sample of the conversion product was isolated and compared with enzyme I from mature bean seeds and the two were shown to possess very similar properties which were quite distinct from those of enzyme II (Table VI). However, there was a difference in carbohydrate content and the values obtained for molecular weights were once again difficult to interpret. Gel filtration studies suggested that both enzymes possessed a molecular weight of about 200 000 but the ultracentrifuge gave a value of 118 000 with the conversion product (which appeared to be homogeneous) and 160 000 with enzyme I from mature seeds. It is again suggested that this anomaly may be due to an association phenomenon and/or differences in carbohydrate content (P. M. Dey, A. Khaleque and J. B. Pridham, unpublished results).

C. CHANGES IN α-GALACTOSIDASES DURING SEED GERMINATION

The early stages of germination of *Vicia faba* seeds approximates to a reversal of maturation as far as the α-galactosidase pattern is concerned. Over the first 12 h of germination there is rapid fall in total and specific α-galactosidase activities and then, after 24 h there are increases to about 50% the original values: no marked changes then occur up to 6 days (Table VII). The Sephadex gel patterns over this period (Fig. 3) show first a rapid decline in enzyme I and then a slow increase in the level of enzyme II.

In the presence of cycloheximide the patterns start to change as in normal germination and then tend to revert to the mature seed pattern with an increase in the activity of enzyme I (Fig. 3) (Dey *et al.*, 1973). Protein synthesis is, therefore, presumed to be involved in germinating seeds in the changes in the levels of the multimolecular forms and when this is interrupted with cycloheximide the increase in enzyme I may arise by a thermodynamically favourable association of units of II, as observed *in vitro*.

Examination of 8-day-old shoots of *Vicia faba* have recently shown that a further form of α-galactosidase, enzyme III, may also exist. Gel filtration suggests that it is a relatively small molecule (mol wt ~38 000) which can be clearly separated from the other isozymes by CM-cellulose chromatography (Fig. 4; P. M. Dey, N. Desai and J. B. Pridham, unpublished results).

TABLE VII

α-Galactosidase activity in germinating *V. faba* seeds (P. M. Dey, P. R. Palan and J. B. Pridham, unpublished results)

Stage of development	Total vol. of extract (ml)	Enzyme activity (mUnit^a/ml)	Total Enzyme activity (mUnit)	Protein (mg/ml)	Specific activity $\left(\dfrac{\text{mUnit/ml}}{\text{mg protein/ml}}\right)$
Mature	30	391	11 730	18·5	21·1
Germination period (h)					
12	30·0	132·3	3 967	12·8	10·3
24	49·0	138·0	6 762	14·0	9·9
48	44·0	155·3	6 831	19·3	8·1
72	47·0	161·0	7 567	20·5	7·9
96	42·0	149·5	6 279	18·0	8·3
120	42·0	143·7	5 937	16·3	8·8
144	42·0	143·7	5 937	17·5	8·2

^{a} One unit of activity is defined as the number of μmoles of *p*-nitrophenyl α-D-galactoside hydrolysed per min at 30°C, pH 5·5.

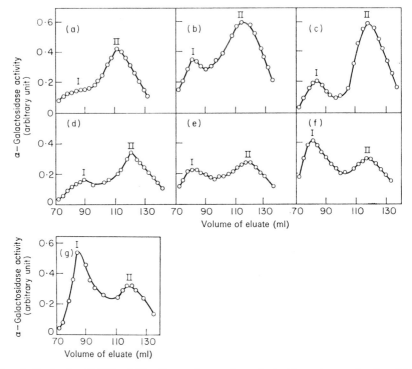

FIG. 3. Sephadex G-100 gel-filtration patterns of α-galactosidases in germinating *Vicia faba* seeds showing the effect of cycloheximide. (a), (b) and (c) represent germination in water after 24, 48 and 72 h, respectively. (d), (e) and (f) show the patterns after corresponding times in the presence of cycloheximide (1 mg/ml) (from Dey *et al.*, 1973).

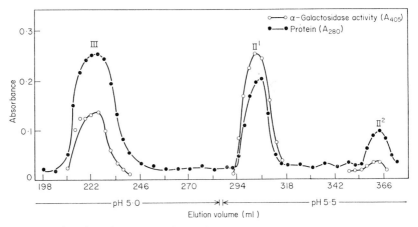

FIG. 4. Fractionation of α-galactosidase II from 8-day-old *Vicia faba* shoots on a CM-52 cellulose column showing enzymes III, II^1 and II^2. (P. M. Dey, N. Desai and J. B. Pridham, unpublished results.)

IV. PHYSIOLOGICAL SIGNIFICANCE OF α-GALACTOSIDASES

Galactose-containing oligo- and poly-saccharides are common carbohydrate reserves in higher plant tissues, particularly in seeds, and there is now little doubt that α-galactosidases function as hydrolytic agents in the utilization of these compounds as sources of energy and cell metabolites. The enzymes may also play an important role in the metabolism of galactolipids (cf. Sastry and Kates, 1964) and in the function of chloroplast membranes (Bamberger and Park, 1966; Gatt and Baker, 1970). Maturation of seeds of *Vicia faba* and other species is a period when the oligosaccharide reserves such as raffinose and stachyose accumulate, together with α-galactosidases. In the case of the broad bean, the mature seed tissues possess the highest levels of activity which may result from the association of units of α-galactosidase II (low specific activity) to form I (high specific activity). In this state a maximum rate of hydrolysis of the reserves can be achieved during the first few hours of germination. If enzyme I is produced from II, it is interesting to speculate regarding the possible controlling influence of K^+ ions on this conversion: they could be involved in simple activation of both enzymes and/or the association of units of II. Germination of *Vicia faba* seeds results in a rapid loss of the high specific activity enzyme, I, and then a gradual increase in the level of II; however, the total activity after 6 days is still very much lower than the original activity of the mature seeds. It is possible that at this time starch takes over as the major reserve of the cotyledons.

ACKNOWLEDGEMENTS

We are indebted to the Science Research Council for financial support and to Dr S. P. Spragg for molecular weight determinations.

We also thank the Central Research Fund of the University of London for an equipment grant.

REFERENCES

Bamberger, E. S. and Park, R. B. (1966). *Pl. Physiol.* **41**, 1591.
Barham, D., Dey, P. M., Griffiths, D. and Pridham, J. B. (1971). *Phytochemistry* **10**, 1759.
Bau, A. (1895). *Chem. Zig. Chem. App.* **19**, 1873.
Courtois, J. E. and Petek, F. (1966). *Meth. Enzym.* **8**, 565.
Courtois, J. E., Petek, F. and Dong, T. (1963). *Bull. Soc. chim. Biol.* **45**, 95.
Dey, P. M., Khaleque, A., Palan, P. R. and Pridham, J. B. (1973). *Biochem. Soc. Trans.* **1**, 661.
Dey, P. M., Khaleque, A. and Pridham, J. B. (1971). *Biochem. J.* **124**, 27P.
Dey, P. M. and Pridham, J. B. (1968). *Phytochemistry* **7**, 1737.
Dey, P. M. and Pridham, J. B. (1969a). *Biochem. J.* **113**, 49.

Dey, P. M. and Pridham, J. B. (1969b). *Biochem. J.* **115**, 47.
Dey, P. M. and Pridham, J. B. (1972). *Adv. Enzymol.* **36**, 91.
Fischer, E. and Lindner, P. (1895). *Ber. Bunsenges. Phys. Chem.* **28**, 3034.
Gatt, S. and Baker, E. A. (1970). *Biochem. biophys. Acta* **206**, 125.
Lee, Y. C. and Wacek, V. (1970). *Archs Biochem. Biophys.* **138**, 264.
Petek, F. and Dong, T. (1961). *Enzymologia* **23**, 133.
Petek, F., Villarroya, E. and Courtois, J. E. (1969). *Eur. J. Biochem.* **8**, 395.
Pridham, J. B. (1958). *Nature, Lond.* **182**, 1687.
Sastry, P. S. and Kates, M. (1964). *Biochemistry* **3**, 1271.
Suzuki, H., Li, Su-Chen and Li, Yu-Teh (1970). *J. biol. Chem.* **245**, 781.
Tanner, W. and Kandler, O. (1968). *Eur. J. Biochem.* **4**, 233.

CHAPTER 7

The Structure and Function of Plant Glycolipids

B. W. NICHOLS

Unilever Research Laboratories, Welwyn, Hertfordshire, England

I. CLASSIFICATION AND STRUCTURE

Although numerous classes of glycolipid, having widely varying structures, are found in higher plants and algae, many of them seldom represent more than a small proportion of the lipophilic constituents of the cell.

The most abundant plant glycolipids are the glycosyl diglycerides, which may be considered as derivatives of 1,2-diglycerides in which a sugar residue is bound glycosidically to the free hydroxyl of the glyceride; members of this group therefore bear structural relationships to the other major classes of lipid, triglycerides and phosphoglycerides (phospholipids).

A. GALACTOSYL DIGLYCERIDES

The first glycosyl diglycerides to be characterized in detail were the mono- and di-galactosyl diglycerides (Figs 1a and 1b) which were obtained by Carter and co-workers (1961a,b) from wheat flour extracts; it was subsequently demonstrated that these compounds are major components of the lipids of leaves and algae (Weenink, 1959). Evidence that some plant tissues may contain small amounts of a trigalactosyl diglyceride has been quoted by various groups and this lipid was more completely characterized by Galliard (1969) who showed that it comprises a digalactosyl diglyceride in which an additional D-galactopyranosyl moiety is linked α-$(1 \rightarrow 6)$ to the terminal galactose unit. Heinz and Tulloch (1969) have shown that some plant extracts contain monogalactosyl diglycerides in which a fatty acyl group is esterified to the hydroxyl group on C-6 of the sugar; this class of compound is possibly only an artefact generated during cellular disruption (Heinz, 1967).

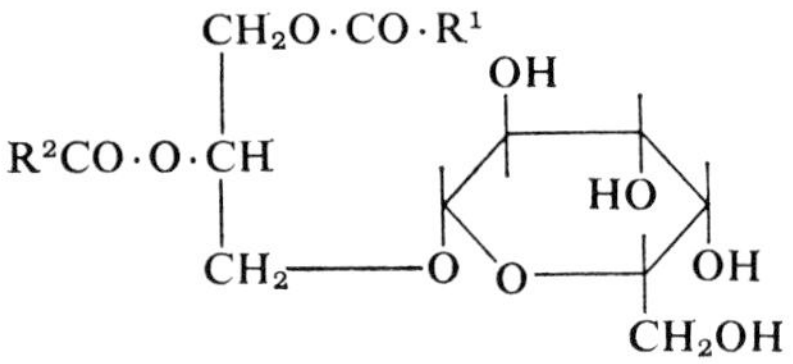

(a) Monogalactosyl diglyceride (1,2-diacyl 3-β-D-galactopyranosyl-L-glycerol)

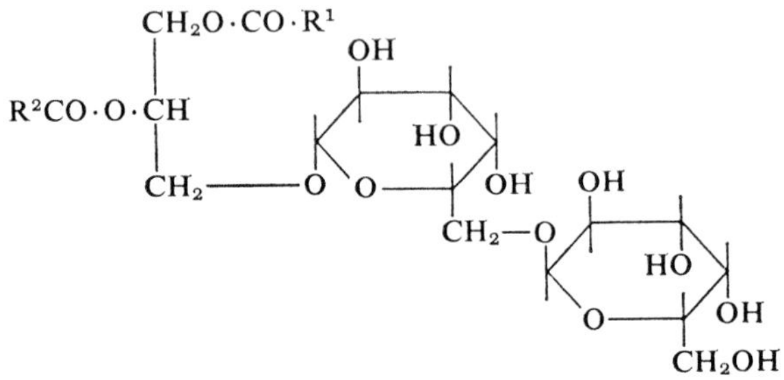

(b) Digalactosyl diglyceride (1,2-diacyl-3-(α-D-galacto-pyranosyl-1,6-β-D-galactopyranosyl)-L-glycerol)

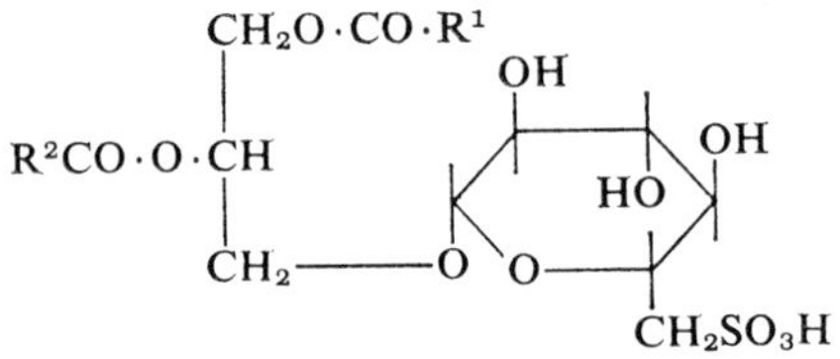

(c) Sulphoquinovosyl diglyceride (1,2-diacyl-3-(6-sulpho-α-D-quinovopyranosyl)-L-glycerol)

FIG. 1. The major glycosyl diglycerides of plants (where R^1COOH, R^2COOH, etc. are fatty acids).

B. SULPHOQUINOVOSYL DIGLYCERIDES

Another class of glycosyl diglyceride, which is apparently also ubiquitous in photosynthetic tissue, is the so-called plant sulpholipid which Benson and co-workers (Benson, 1963) showed to be a diglyceride derivative of 6-deoxyglucose-6-sulphonic acid (Fig. 1c).

C. OTHER GLYCOLIPIDS

The particular classes of glycosyl diglyceride referred to above are almost exclusive to the vegetable kingdom, although small quantities infrequently occur in animals and a few bacteria. Indeed, the only general class of glyco-lipid found in substantial quantities in both animal and plant tissues is cerebroside (Fig. 2a) in which a hexose residue is attached glycosidically to the terminal hydroxyl group of a long chain fatty amino alcohol (sphingo-sine); the amino group of the sphingosine is amidically combined with fatty acid. In plant cerebrosides the hexose residue is invariably glucose, although the structures of the fatty alcohols and acids may vary widely (Hitchcock and Nichols, 1971). These compounds are apparently confined to the tissues of higher plants and have not been detected in algae; a similar specificity in distribution within the plant kingdom holds for the steryl glucosides (Fig. 2b) in which a glucose residue is attached to sterol. Esterified sterol glucosides, in which fatty acid residues are esterified to the 6-hydroxyl group of the

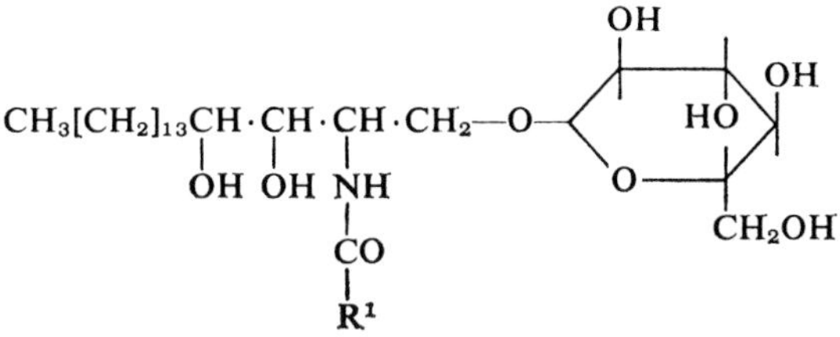

(a)

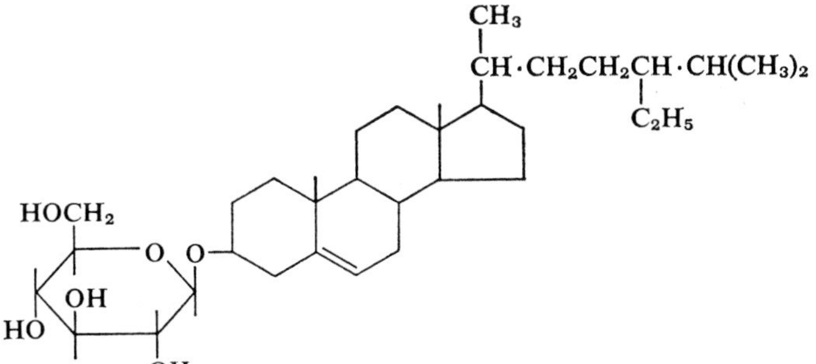

(b)

FIG. 2. Cerebroside (a) and sterol glucoside (b).

glucose, are also widely distributed amongst the tissues of higher plants (Lepage, 1964).

The most complex class of glycolipid yet isolated from plants is the phytoglycolipid which was originally isolated by Carter and co-workers (1958) from the phospholipid fractions of a number of oilseeds. These complex compounds contain fatty acids, phosphorus, pentoses, hexoses, amino sugars, uronic acids, inositol and long chain bases of the sphingosine type. For the fraction from *Zea mays* Carter and co-workers (1969) have proposed the partial structure illustrated in Fig. 3a, in which the point of attachment and sequence of the terminal monosaccharides remain undefined.

Wagner and co-workers (1969a) isolated a related fraction from peanut phospholipids which differed from that of maize in that it contained no fucose. These workers proposed the structure illustrated in Fig. 3b for their preparation which differs from that of Carter *et al.* in the site of attachment of the mannose moiety; little evidence in support of this difference has been quoted, however. It is likely that this peanut phytoglycolipid was a major

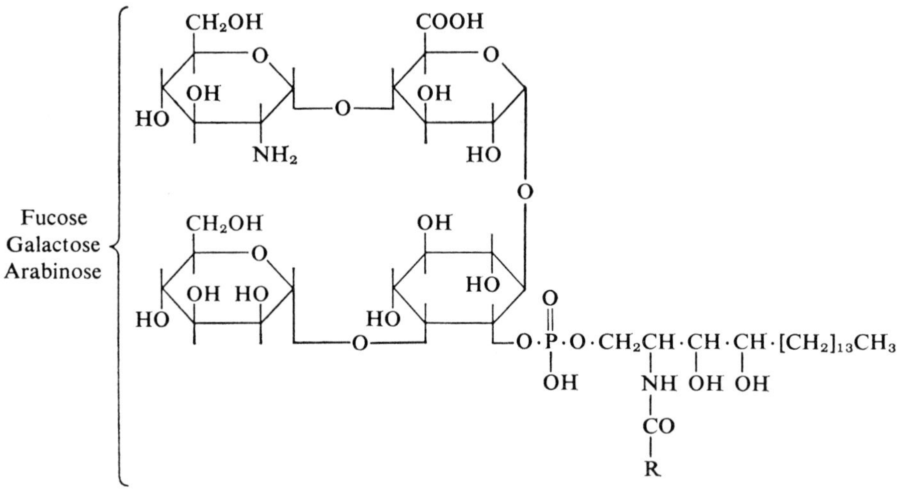

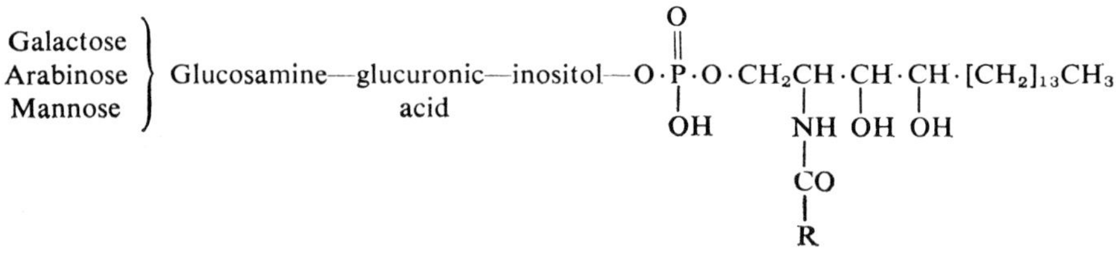

FIG. 3. Proposed structures for complex sphingoglycolipids of plants.

constituent of the complex lipid which was isolated and partially characterized by Malkin and Poole in 1953.

The experimental procedures employed for the preparation of these phyto-glycolipid fractions have usually involved mild alkaline hydrolysis of crude phospholipid fractions as a preliminary step, which raises the possibility that in the plant cell the phytoglycolipid may be part of an even more complex lipid. Comparison between the countercurrent distribution patterns of the purified lipid and that of the crude fraction from which the lipid is obtained by hydrolysis (Carter *et al.*, 1962) suggests that this could be the case.

Although almost all of the earlier work on the isolation and characteriza-tion of phytoglycolipid was carried out on oilseed fractions, Carter and Koob (1969) isolated a phytoglycolipid from bean leaves which differed from similar preparations from flax and maize seed, and resembled that from peanut, in containing no fucose.

A preliminary note by Wagner and co-workers (1969b) describes the partial characterization of a lipid from *Scenedesmus obliquus* which resembles the phytoglycolipids in many respects but does not contain inositol. This is the first indication that algae synthesize sphingosines or sphingolipids.

Another class of sphingoglycolipid, as yet also incompletely characterized, is one isolated from crude oilseed inositol lipid fractions and which differs from phytoglycolipid in containing no amino sugar (Carter and Kisic, 1969).

D. HETEROCYST GLYCOLIPIDS

Glycolipids apparently specific to nitrogen fixing (heterocystous) algae were originally isolated and partially characterized by Nichols and Wood (1968) and Bryce *et al.* (1972). It was demonstrated that these compounds are monohexoside derivatives of a series of long chain polyhydroxy alcohols, the most important in *Anabaena cylindrica* being 1,3,25-trihydroxyhexa-

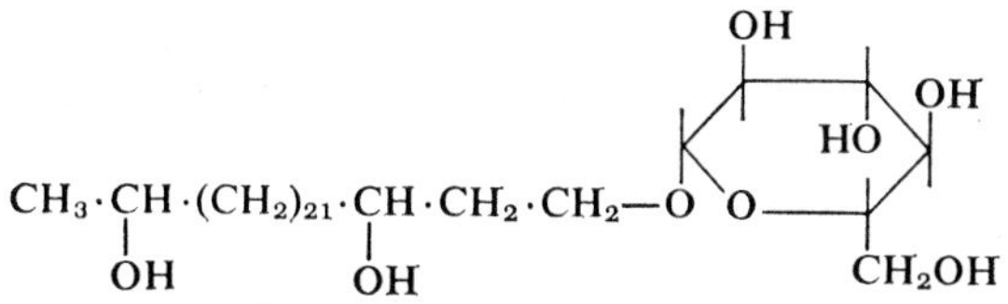

(a) 1-(O-α-D-glucopyranosyl)-3,25-hexacosanediol

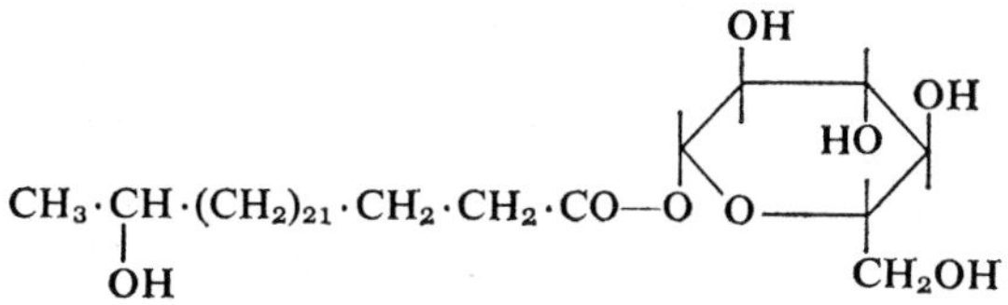

(b) 25-hydroxyhexacosanoic-(1-α-D-glucopyranose) ester

FIG. 4. Classes of glycolipid isolated from *Anabaena cylindrica*.

cosane. The same workers found that the major hexoses in these lipids were glucose and galactose, and subsequent studies by Lambein and Wolk (1973) on extracts from the same algae showed that these hexoses are attached to the primary hydroxyl group (Fig. 4a). Lambein and Wolk also found that extracts of *Anabaena* contained similar derivatives of a C_{28} polyhydroxy alcohol and, in addition, C_{26} and C_{28} carboxyl glycosylated, hydroxy fatty acids (Fig. 4b) were also identified.

Thus these four unique lipids in *A. cylindrica* are structurally related in that they contain long straight hydrocarbon chains with glucose or galactose residues at one end and free ω-1-hydroxyls at the other.

E. MONOTERPENE DERIVATIVES

Monoterpene β-D-glucosides represent a major proportion of the total (free and combined) monoterpene content of certain tissues, especially rose petals (Francis and Allcock, 1969); such glycosides do not entirely fit into the common definition of glycolipid because they are freely soluble in water.

Those glycolipids which contain fatty acid residues normally occur as a series of molecular species of similar general composition but containing a variety of fatty acid moieties of different structure. A description of the nature and significance of these acid residues would be out of place here, but was discussed in a recent monograph (Hitchcock and Nichols, 1971).

II. DISTRIBUTION

By far the most abundant and widely distributed plant glycolipids are the galactosyl diglycerides and sulphoquinovosyl diglycerides, which are ubiquitous constituents of photosynthetic lamellae and are consequently major lipid components of leaves and algae (Table I). Because such lamellae are

TABLE I

Glycerolipid composition of photosynthetic tissues
(from Roughan and Batt, 1969)[a]

Plant	Galactosyl diglycerides		Sulpholipid	Total phospholipid
	MGD[b]	DGD[c]		
Moss (unidentified)	2·68	1·50	0·48	2·26
Maidenhair tree (*Ginkgo biloba*)	4·70	2·80	0·30	3·72
White clover (*Trifolium repens*)	8·60	5·20	0·76	2·94
Tomato (*Solanum esculentum*)	5·08	2·46	0·31	2·18
Lettuce (*Lactuca sativa*)	0·68	0·68	0·03	0·81
Perennial ryegrass (*Lolium perenne*)	5·10	3·95	0·45	3·15
Maize (*Zea mays*)	3·10	2·30	0·35	1·53

[a] μmol lipid per g fresh weight.
[b] Monogalactosyl diglyceride.
[c] Digalactosyl diglyceride.

TABLE II

Lipid composition of avocado and cauliflower mitochondria and chloroplasts
(from Schwertner and Biale, 1973)

	Avocado		Cauliflower	
	Mitochondria 38	Chloroplasts 38	Mitochondria 33	Chloroplasts 34
Total lipid (%)[a]				
	Weight % of total lipid[b]			
Phospholipid	50·0	20·4	54·0	9·1
Monogalactosyl diglyceride	2·1	9·0	2·2	19·4
Digalactosyl diglyceride	3·0	6·7	2·7	7·4
Sulpholipid	0·9	3·6	1·3	2·5
Chlorophyll	0·04	8·2	0·03	33·6
Fatty acids of total neutral lipid	16·5	ND	18·9	18·2
	nmol/mg protein N			
Phospholipid	2455	995	2467	365
Monogalactosyl diglyceride	101	360	99	777
Digalactosyl diglyceride	122	223	102	248
Sulpholipid	39	135	56	95
Chlorophyll	2	286	2	1172
Fatty acids of total neutral lipid	2394	ND	2265	1922

ND, not determined.

[a] Total lipid = (lipid/lipid + protein) × 100.

[b] Nanomoles of lipid were converted to weight using the following calculated molecular weights: monogalactosyl diglyceride, 786·6; digalactosyl diglyceride, 948·6; sulpholipid, 837·0; phospholipid, 784·3; chlorophyll, 905·0.

major structural features of the blue-green algae, these lipids are particularly predominant in this class of organism.

Although chloroplasts are undoubtedly the major site of accumulation of the glycosyl diglycerides, these lipids also occur in other plant organelles such as mitochondria and endoplasmic reticulum. Recent work by Schwertner and Biale (1973) has shown that plant mitochondria contain significant quantities of all three glycolipids even though their abundance, relative to phospholipid, is much lower than that found in chloroplasts (Table II).

The galactosyl glycerides have also been shown to comprise a major part of the reserve fat in the seeds of some *Briza* (quaking grass) species; the seeds of *Briza spicata*, for example, contain 20% of lipid of which 49% is digalactosyl diglyceride and another 29% is monogalactosyl diglyceride (Smith and Wolff, 1966). In these cases the organelles from which the lipids originate have not been identified; in most seeds the reserve fat (triglycerides) occurs in so-called "fat bodies" (Harwood *et al.*, 1971) but the structure of *Briza* seeds has not been reported. Such high levels of glycolipids in seeds are most uncommon and even amongst *Briza* species they represent the exception rather than the rule.

Although the major glycolipids in most blue-green algae are the glycosyl glycerides, large proportions of monohexoside derivatives of long chain polyhydroxy alcohols and hydroxy acids may occur in nitrogen-fixing algae. These lipids only occur when the alga has fixed nitrogen during growth, and they appear to be specific to the specialized cells (heterocysts) which are formed only during nitrogen fixation (Walsby and Nichols, 1969).

The sites of accumulation of the other classes of plant glycolipid have still to be identified. The most abundant and ubiquitous of these, the sterol glucosides and their 6-*O*-acyl esters, seem to be virtually absent from chloroplasts, mitochondria and endoplasmic reticulum, and are not found at all in algae. A possible site of accumulation of these compounds would, therefore, appear to be the cell wall of higher plants.

III. BIOGENESIS

A. GLYCOSYL DIGLYCERIDES

The biosynthetic pathways leading to the formation of the galactosyl glycerides has now been studied in a number of systems. In most cases the general pathway appears to be similar, namely that the monogalactosyl diglyceride is synthesized by reaction between 1,2-diglyceride and UDP-galactose. UDP-galactose is also required for the synthesis of digalactosyl diglyceride in a reaction in which monogalactosyl diglyceride, rather than diglyceride, is the preferred acceptor (Ongun and Mudd, 1968). The same workers also obtained evidence that plants catalyse the transfer of mono-saccharide from UDP-galactose to di- and tri-galactosyl diglycerides with the formation of tri- and tetra-galactosyl diglycerides, respectively. Thus the general pathway of galactosyl diglyceride synthesis in plants is probably:

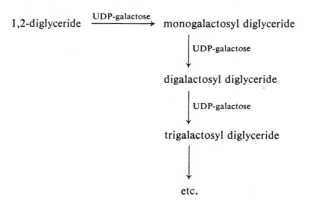

By analogy with the biosynthetic pathways proposed for the galactosyl diglycerides, Benson (1963) suggested that the plant sulpholipid might be synthesized by transfer to diglyceride of the sulphoquinovose group of a nucleoside diphosphosulphoquinovose, which has been identified in extracts of *Chlorella vulgaris* by Zill and Cheniae (1962). As yet, no experimental data have been offered in support of this or any other mechanism for the biogenesis of the sulpholipid, although several authors have speculated on the biological origins of the sulpho-sugar (Benson, 1963; Zill and Cheniae, 1962; Davies *et al.*, 1966).

B. STEROL GLYCOSIDES

In studies employing cellular and subcellular systems from a variety of plant tissues, Ongun and Mudd (1970) obtained synthesis of sterol glycosides and acylated sterol glycosides when either UDP-glucose or UDP-galactose were supplied as precursors. The time course of the incorporations indicated sterol glycoside to be the precursor of the acylated sterol glycoside.

IV. FUNCTIONS

A. GLYCOSYL DIGLYCERIDES

Few physiological functions for the plant glycolipids have been unequivocally established, although many have been proposed. There is no doubt, however, that the glycosyl diglycerides play an important role in the maintenance of thylakoid membrane structure in chloroplasts, and Weier and Benson (1967) and Branton and Park (1967), and others, have suggested the manner in which these lipids are oriented with respect to protein and other membrane components. Most modern theories now assume that protein–lipid association in the thylakoid occurs by hydrophobic association of protein with the acyl residues of the glycolipids; this implies that the sugar moieties will be interacting with the aqueous phase of the membrane system.

In studies of lipid synthesis in greening cultures of *Euglena gracilis*, Rosenberg and Pecker (1964) noted that the sulpholipid accumulated before measurable quantities of chlorophyll and that subsequently the rate of synthesis of the glycolipid closely paralleled that of chlorophyll. They therefore concluded that the sulpholipid might be required for the orientation and functioning of chlorophyll. Studies of chlorophyll–sulpholipid interactions *in vitro* seem to support this theory (Trosper and Sauer, 1968).

An interesting proposal for the mechanism of sugar transport through membranes was advanced by Benson (1963) on the basis of studies in which the sugar moieties of algal galactosyl diglycerides showed rapid turnover of label following incubation of the organism with $^{14}CO_2$ (Ferrari and Benson, 1961). In this theory glucose epimerizes to galactose which reacts with diglyceride giving galactosyl diglycerides, in which form the sugar is moved across the membrane and is then released by β-galactosidase activity.

Some doubt whether this mechanism can be an important one in photosynthetic tissues has been cast by subsequent studies by Chang and Kulkarni (1970) and Roughan (1970) who were unable to detect significant turnover of galactose residues in the galactosyl diglycerides of spinach and pumpkin leaves.

Another function of the galactosyl glycerides may be that of energy storage; this was originally proposed by Rosenberg (1967) on the basis of observations that the levels of these compounds diminished precipitously when healthy green cells of *Euglena gracilis* were starved of light and nutrients, until a ratio of one molecule of chlorophyll to two molecules of galactosyl diglyceride was reached. Beyond this point chlorophyll and galactosyl diglycerides disappeared together and the organism became dormant and disintegrated. These glycolipids may also serve as energy reserves in those seeds such as *Briza spicata* which contain high galactosyl diglyceride/triglyceride ratios. Miyachi and Miyachi (1966) have suggested that in some situations sulpholipid may serve as an emergency source of carbon and sulphur.

Nichols and co-workers (e.g. Safford and Nichols, 1970) followed the course of ^{14}C-label through the component fatty acids of monogalactosyl diglyceride fractions of algae and concluded that this lipid is intimately involved in the biosynthesis of polyunsaturated acids in algae.

B. HETEROCYST GLYCOLIPIDS

The glycolipids which appear to be specific to heterocyst forming blue-green algae are localized in the laminated layer of the heterocyst envelope. Winkenbach *et al.* (1972) have pointed out that because this layer is hydrophobic and possibly crystalline it may serve as a barrier to the movement of water and hydrophilic molecules between the protoplast of the heterocyst and the growth medium; it may be that such compounds enter and leave heterocysts via the constricted glycolipid-lined channel at each heterocyst pore.

C. OTHER FUNCTIONS

Other functions proposed for plant glycolipids, and an account of their general metabolic behaviour, are contained in a recent monograph (Hitchcock and Nichols, 1971).

REFERENCES

Benson, A. A. (1963). *Adv. Lipid Res.* **1**, 387.
Branton, D. and Park, R. B. (1967). *J. Ultrastruct. Res.* **19**, 283.
Bryce, T. A., Welti, D., Walsby, A. E. and Nichols, B. W. (1972). *Phytochemistry* **11**, 295.
Carter, H. E., Gigg, R. H., Law, J. H., Nakayama, T. and Weber, E. (1958). *J. biol. Chem.* **233**, 1309.
Carter, H. E., Ohno, K., Nojima, S., Tipton, C. L. and Stanacev, N. Z. (1961a). *J. Lipid Res.* **2**, 214.
Carter, H. E., Hendry, R. A. and Stanacev, N. Z. (1961b). *J. Lipid Res.* **2**, 223.
Carter, H. E., Galanos, D. S., Hendrickson, H. S., Jann, B., Nakayama, T., Nakaxawa, Y. and Nichols, B. W. (1962). *J. Am. Oil Chem. Soc.* **39**, 107.
Carter, H. E., Strobach, D. R. and Hawthorne, J. R. (1969). *Biochemistry* **8**, 383.
Carter, H. E. and Kisic, A. (1969). *J. Lipid Res.* **10**, 356.
Carter, H. E. and Koob, J. L. (1969). *J. Lipid Res.* **10**, 363.
Chang, S. B. and Kulkarni, N. D. (1970). *Phytochemistry* **9**, 927.
Davies, W. H., Mercer, E. I. and Goodwin, T. W. (1966). *Biochem. J.* **98**, 369.
Ferrari, R. A. and Benson, A. A. (1961). *Archs Biochem. Biophys.* **93**, 185.
Francis, M. J. O. and Allcock, C. (1969). *Phytochemistry* **8**, 1339.
Galliard, T. (1969). *Biochem. J.* **115**, 335.
Harwood, J. L., Sodja, A., Stumpf, P. K. and Spurr, A. R. (1971). *Lipids* **6**, 851.
Heinz, E. (1967). *Biochim. biophys. Acta* **144**, 333.
Heinz, E. and Tulloch, A. P. (1969). *Hoppe-Seyler's Z. physiol. Chem.* **350**, 493.
Hitchcock, C. H. and Nichols, B. W. (1971). "Plant Lipid Biochemistry", pp. 54–55. Academic Press, London and New York.
Lambein, F. and Wolk, C. P. (1973). *Biochemistry* **12**, 791.
Lepage, M. (1964). *J. Lipid Res.* **5**, 587.
Malkin, T. and Poole, A. G. (1953). *J. chem. Soc.* 3470.
Miyachi, S. and Miyachi, S. (1966). *Pl. Physiol.* **41**, 479.
Nichols, B. W. and Wood, B. J. B. (1968). *Nature, Lond.* **217**, 767.
Ongun, A. and Mudd, J. B. (1968). *J. biol. Chem.* **243**, 1558.
Ongun, A. and Mudd, J. B. (1970). *Pl. Physiol.* **45**, 255.
Rosenberg, A. (1967). *Science, N.Y.* **157**, 1192.
Rosenberg, A. and Pecker, M. M. (1964). *Biochemistry* **3**, 254.
Roughan, P. G. (1970). *Biochem. J.* **117**, 1.
Roughan, P. G. and Batt, R. D. (1969). *Phytochemistry* **8**, 363.
Safford, R. and Nichols, B. W. (1970). *Biochim. biophys. Acta* **210**, 57.
Schwertner, H. A. and Biale, J. B. (1973). *J. Lipid Res.* **14**, 235.

Smith, C. R. and Wolff, I. A. (1966). *Lipids* **1**, 123.

Trosper, T. and Sauer, K. (1968). *Biochim. biophys. Acta* **162**, 97.

Wagner, J., Zofcsik, W. and Heng, I. (1969a). *Z. Naturf.* **246**, 922.

Wagner, H., Pohl, P. and Munzing, A. (1969b). *Z. Naturf.* **246**, 360.

Walsby, A. E. and Nichols, B. W. (1969). *Nature, Lond.* **221**, 673.

Weenink, R. O. (1959). *N.Z. Jl Sci.* **2**, 273.

Weier, T. E. and Benson, A. A. (1967). *Am. J. Bot.* **54**, 389.

Winkenbach, F., Wolk, C. P. and Jost, M. (1972). *Planta* **107**, 69.

Zill, L. P. and Cheniae, G. (1962). *A. Rev. Pl. Physiol.* **13**, 225.

CHAPTER 8

Some Aspects of the Enzymic Degradation of Starch*

D. J. MANNERS

*Department of Brewing and Biological Sciences,
Heriot-Watt University, Edinburgh, Scotland*

I. INTRODUCTION

Studies on various aspects of the enzymic degradation of starch are continuing in many laboratories throughout the world. The combined efforts of chemists, biochemists, microbiologists and plant physiologists result in the appearance of literally hundreds of publications each year, and it is clearly quite impracticable to attempt to give a comprehensive review of even the most recent developments. This review will therefore be confined to two aspects of the subject which form part of the special research interests of this Department, namely, the possible existence of multiple forms of starch-degrading enzymes and the changes which occur in starch during the germination of cereals.

A substantial proportion of the earlier work is described in review articles (e.g. Manners, 1962; Greenwood and Milne, 1968; Robyt and Whelan, 1968) which provide the basic information on the occurrence, isolation and properties of the starch degrading enzymes.

---

* This review is dedicated to the memory of Dr David James Bell (1905–1972), who firsti ntroduced the writer to the enzymic degradation of polysaccharides.

II. THE EXISTENCE OF MULTIPLE FORMS OF STARCH-DEGRADING ENZYMES

A. INTRODUCTION

1. *Isoenzymes*

An important development in enzyme chemistry during the last decade has been the demonstration that many enzymes exist in multiple forms, i.e. as isoenzymes (isozymes). This latter term was originally used by Markert and Møller (1959) to describe different molecular forms of a protein which showed the same enzyme specificity; zone electrophoresis was used to detect these isoenzymes. More recently, the term has been used to describe a family of enzymes, from the same biological source, which have the same general specificity, but differ in the details of their physical and catalytic properties. The current literature contains numerous references to reports of amylase isoenzymes; indeed, the detection of so-called "isoenzymes" by electrophoresis of various plant protein preparations has become something of a fashionable occupation. However, the detection of protein bands on a gel after electrophoresis is not necesssarily definite evidence for the presence of isoenzymes and many of the results described may, in fact, be due to artefacts.

Certain enzymes, e.g. phosphoglucose isomerase, exist in oxidized and reduced forms and when subjected to electrophoresis in the absence of reducing agents may give multiple bands; multiple peaks may also be observed on isoelectric focusing in sucrose density gradients. When these experiments are repeated in the presence of a reducing agent such as 2-mercaptoethanol or dithiothreitol, only a single peak is observed (Payne *et al.*, 1972), suggesting that previous reports of isoenzymes of phosphoglucose isomerase describe artefacts arising from the partial oxidation of the enzyme. A similar situation exists with β-amylase, which is well known as a thiol-enzyme; purified enzyme preparations which were homogeneous on electrophoresis at pH 4·5 produced two separate enzyme bands at pH 8·9 (LaBerge and Meredith, 1971). This change was attributed to instability of the enzyme at alkaline pH values caused by oxidation of essential thiol-groups.

Multiple forms of a number of α-amylases have been reported. Banks *et al.* (1971) have shown that crystalline porcine pancreatic α-amylase can be fractionated into four components by ion-exchange chromatography. However, the molecular weights and amino acid compositions of the four forms were identical. Since $\sim 20\%$ of all the amino acid residues present were either aspartic or glutamic acid, which could occur either as the free acid or as the amide, it was suggested that conversion of a proportion of the amide groups into the free acid could occur during isolation and purification of the enzyme, giving proteins which would then differ in charge and become separable by ion-exchange chromatography. Hence, the four apparent multiple forms could also be artefacts.

Finally, it should be noted that the number of isoenzymes apparently present in unheated extracts of malted barley was decreased by heat treatment (Frydenberg and Nielsen, 1965) suggesting that at least some of the "isoenzymes" were unstable forms of the same protein.

In the following review, observations of multiple bands obtained by the electrophoresis of various starch-degrading enzymes will not be regarded as definitive evidence for the existence of isoenzymes, unless it is supported by additional evidence based on protein fractionation and characterization. For this reason, the following account may appear not to be comprehensive, but at least discussion of experimental results which may be artefactual will be avoided. This is important in view of the genetic implications of the existence of isoenzymes.

2. Starch-metabolizing enzymes

Present knowledge of the major enzymes involved in the degradation of the linear and branched components of starch is summarized in Fig. 1. It should be emphasized that of the five types of enzyme shown, only the α-amylases are believed to be able to degrade intact starch granules. Hence, in vivo, some degree of α-amylolysis is a necessary prerequisite for starch degradation by either hydrolytic or phosphorolytic enzyme systems. For in vitro studies, the granules can be solubilized and the two starch components separated and isolated, and the actions of single purified enzymes then examined. In

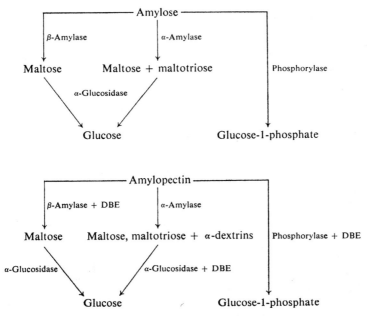

FIG. 1. Pathways for the degradation of the starch components (DBE = Debranching enzyme).

extrapolating these latter types of result to the *in vivo* situation, it is important to realize that the two systems are not always strictly comparable.

In addition to the above enzymes, Jørgensen (1964) has isolated an α-glucosidase from malted barley which hydrolyses maltose, isomaltose and panose, and also acts directly on soluble starch, releasing glucose as a primary product of enzyme action. This enzyme also hydrolyses dextran and amylopectin β-limit dextrin so that enzyme action is not arrested by α-(1 → 6)-glucosidic linkages. In mammalian tissues, an α-glucosidase (γ-amylase) may be involved in the direct conversion of glycogen into glucose. Whether the malted barley enzyme plays a similar *in vivo* role with starch is not yet known.

B. β-AMYLASES

This group of enzymes, which occur only in the cereals and in certain other higher plants such as sweet potatoes and soya beans, catalyse a stepwise hydrolysis of alternate linkages in starch-type polysaccharides with the liberation of maltose. Enzyme action on linear substrates is usually complete; with branched substrates, the enzyme is unable to either hydrolyse or by-pass the α-(1 → 6)-D-glucosidic inter-chain linkages so that enzyme action is incomplete, the products being maltose and a β-limit dextrin. With most samples of amylopectin, the percentage conversion to maltose (i.e. β-amylolysis limit) is 50–60.

Ungerminated barley is one of the normal sources of β-amylase. Part of the enzyme is easily extracted with dilute solutions of salts or buffers, and is often referred to as free, soluble or active β-amylase. The remainder of the enzyme becomes solubilized after treatment of the cereal with papain or reducing agents, and hence is termed bound, insoluble or latent β-amylase. LaBerge and Meredith (1969) showed that thiol-containing extracts of barley contained two forms of β-amylase which were separable by ion-exchange chromatography. These forms were referred to as β-amylases I and II. During the steeping of barley prior to germination, β-amylase I disappeared and was absent from the corresponding malt extracts. β-Amylase II remained and an enzyme of similar chromatographic properties was present in the final malt extract which was termed β-amylase III. However, the major β-amylase component of malted barley (β-amylase IV) first appeared during the early stages of germination and the amount increased continuously during the remainder of the germination process. The chromatographic behaviour of the β-amylases from malted barley is shown in Fig. 2. These experiments show the real existence of two forms of β-amylase in both ungerminated and germinated barley, only one of which appears to remain unchanged during the germination process.

In wheat, β-amylase also exists in multiple forms. Kruger (1970) examined both the free and latent β-amylases, the latter prepared by treating glutenin with 1-thioglycerol. Ion-exchange chromatographic studies showed that both

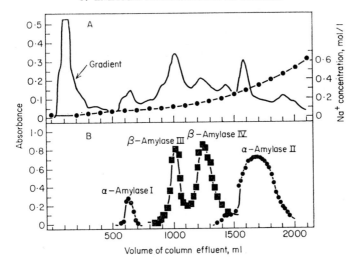

FIG. 2 Fractionation of barley-malt extracts by gradient elution chromatography on carboxymethyl cellulose. In A, solid line = absorbance of column effluent at 280 nm; closed circles = Na⁺ gradient. B = amylase activities. Reproduced with permission from MacGregor *et al.* (1971).

enzyme preparations contained two components. When the first component from the free and latent β-amylases were combined and rechromatographed, only one major protein peak was obtained. A similar result was obtained with the second component.

Although these results show that barley and wheat β-amylases occur in two forms, the chemical differences between them and their respective roles *in vivo* are not yet known. Regarding other common cereals, β-amylase in malted sorghum occurs in only one form (Botes *et al.*, 1967a) but no further information is available. There is limited evidence (see Table I) of heterogeneity of some crystalline preparations of sweet potato β-amylase (Siepmann and Stegemann, 1967) but whether this represents the situation in the actual plant tissues has not been determined. It therefore seems probable that some, but not all, of the β-amylases exist in multiple forms.

C. α-AMYLASES

These enzymes, which are widely distributed in nature, initially catalyse an essentially random hydrolysis of the non-terminal α-(1 → 4)-glucosidic linkages in both linear and branched substrates. As the hydrolysis proceeds, the degradation becomes less random since there is a significant increase in the relative proportion of linkages which are adjacent to an inter-chain linkage, or a terminal glucose residue, for which the enzyme has a much lower affinity. The normal end products are maltose and glucose from amylose and these sugars, together with branched oligosaccharide α-dextrins, from amylopectin. At low enzyme concentrations, maltotriose may also be present.

TABLE I

Existence of multiple forms of starch-degrading enzymes

Enzyme	Source	Number of forms	References
β-Amylase	Ungerminated barley	2	LaBerge and Meredith (1971)
	Germinated barley	2	LaBerge and Meredith (1971)
	Sweet potato	2[a]	Siepmann and Stegemann (1967)
	Malted sorghum	1	Botes, Joubert and Novellie (1967a)
	Wheat	2	Kruger (1970)
α-Amylase	Human parotid saliva	2	Keller et al. (1971)
	Germinated barley	2	LaBerge and Meredith (1971)
	Malted wheat	4	Kruger and Tkachuk (1969)
	Malted rye	2	Manners and Marshall (1972)
	Malted sorghum	1–4 (?)	Botes, Joubert and Novellie (1967b)
α-Glucosidase	Rice	2	Takahashi et al. (1971)
	Sweet corn	3	Marshall and Taylor (1971)
	Barley	1 (?)	Jørgensen (1964)
	Alfalfa	3	Hutson and Manners (1965)
Starch phosphorylase	Potato	3 (?)	Gerbrandy and Doorgeest (1972)
	Vicia faba	2–4 (?)[b]	Gerbrandy and Verleur (1971)
	Phaseolus vulgaris	1–3 (?)[b]	Gerbrandy and Verleur (1971)
	Maize	4	Tsai and Nelson (1969)
	Mistletoe	2	Khanna et al. (1971)
	Spinach	2	Fekete (1968)

[a] One major and one minor zone detected by electrophoresis of a crystalline enzyme preparation.
[b] Results varied with the tissues examined and the maturity of the plant.

There is now good evidence for the existence of multiple forms of human salivary α-amylase which differ with respect to both molecular weight and chemical composition (Keller *et al.*, 1971). Rat pancreatic α-amylase also exists in two forms (Sanders and Rutter, 1972). Both of these observations on mammalian enzymes have implications with respect to *in vitro* studies on starch degradation but the emphasis in this section must be on plant enzymes which may have *in vivo* significance.

In general, ungerminated cereals contain little or no α-amylase and only limited information is available on the properties of these enzymes. However, the germination of cereals results in a dramatic increase in α-amylase activity, and, accordingly, most of the published work deals with the properties of the enzyme from germinated or malted cereals.

α-Amylase from malted Canadian hard red spring wheat has been purified by Kruger and Tkachuk (1969). Four components were detected by chromatography on DEAE-cellulose which, in order of elution from the column, contained 9, 34, 21 and 36% of the total activity which was eluted. The over-all recovery was 27%. This would suggest that malted wheat contains three major and one minor α-amylase component but the precise relationship between them remains to be elucidated.

MacGregor *et al.* (1971) have separated the α-amylase from malted barley into two components (see Fig. 2) which had different chromatographic mobilities. These components were considered not to be artefacts arising from proteolysis or produced by the isolation procedure. The α-amylases I and II were eluted at optimum salt concentrations of 0·04 and 0·14 M respectively and also differed in electrophoretic mobility (MacGregor and Meredith, 1971).

Malted rye also contains two α-amylase components which differ in mobility on continuous curtain electrophoresis, in molecular weight (as shown by gel-filtration on Sephadex G-100), in stability on dialysis and towards EDTA (Manners and Marshall, 1972).

Although there is therefore positive evidence of multiple forms of α-amylase in germinated or malted barley, rye and wheat, sorghum appears to be different. Botes *et al.* (1967b) purified the latter enzyme to homogeneity by physicochemical criteria. During chromatography on a column of calcium phosphate, four sub-fractions were obtained, but since these had almost identical physical, chemical and kinetic properties, these observations were not taken as evidence of isoenzymes and the situation may be similar to that already described for the porcine pancreatic α-amylase preparation (see Banks *et al.*, 1971).

In some laboratories, extracts of cereal seeds and their aleurone layers have been examined electrophoretically for α-amylase isoenzymes (e.g. Jacobsen *et al.*, 1970; Bilderback, 1971; Chao and Scandalios, 1972), and the effect of gibberellic acid on the rate of development of the various forms has also been investigated. The number of bands of activity varied from four to seven, but for the reasons stated in the introduction, it is premature at the present time to draw reliable conclusions as to the number of α-amylases actually present and the physiological processes controlling their development.

D. α-GLUCOSIDASES

The final stage in the enzymic hydrolysis of starch to the monosaccharide level is believed to involve the hydrolysis of the products of amylase action, namely maltose and related oligosaccharides, to glucose. This latter reaction is catalysed by α-glucosidases which appear to be widely distributed in higher plants. Extracts of eleven plant tissues were examined by Hutson and Manners (1965) and shown to hydrolyse not only maltose but also the corresponding α-(1 → 3)- and α-(1 → 6)-linked disaccharides, nigerose and isomaltose. In general, the activity towards maltose was the highest and there was considerable variation in the relative activity towards nigerose and isomaltose. With most plants, the respective nigerase and isomaltase activities were 70–90% and 10–50% of that towards maltose. It seems unlikely that the first two activities have any significance *in vivo* but merely represent facets of a group specific enzyme.

The α-glucosidases in alfalfa were separated by continuous curtain electrophoresis to give three enzyme fractions which showed significant variations in the relative nigerase:isomaltase activity of 78:26, 40:66 and 77:44, respectively (Hutson and Manners, 1965). Although homogeneous enzymes were not obtained, it was suggested that the α-glucosidase might exist as isoenzymes.

As far as plants which actively metabolize starch are concerned, Takahashi *et al.* (1971) have isolated two forms of α-glucosidase (I and II) from rice seeds, both of which were homogeneous by physicochemical criteria. The two forms of the enzyme had different K_m values for maltose and hydrolysed maltotriose and maltotetraose at different rates and showed different sensitivities towards inhibitors such as Tris and erythritol.

α-Glucosidases showing optimum activity at acidic pH values (3·1–3·8) have been isolated from sweetcorn (Marshall and Taylor, 1971). Chromatography on hydroxylapatite gave three fractions which had similar pH activity curves, and gave single bands on electrophoresis, suggesting they were true isoenzymes. α-Glucosidases showing a similarly acidic optimum pH value were also detected in extracts of normal maize, waxy maize and a commercial malt. An α-glucosidase from malted barley has also been purified some 20-fold (Jørgensen and Jørgensen, 1963; Jørgensen, 1963). The latter enzyme hydrolyses maltose, isomaltose and panose and the results suggest that only one enzyme is involved in the hydrolysis of both types of glucosidic linkage. As mentioned previously (p. 112) it also hydrolyses soluble starch.

Overall, the study of plant α-glucosidases has been rather neglected, compared to that of the amylases or phosphorylases, and only a limited amount of information on their characteristics is at present available.

E. STARCH PHOSPHORYLASES

It is now more than 30 years since Hanes (1940) first reported the presence of phosphorylase in preparations from peas and potatoes, which catalysed a reversible reaction between α-D-glucose-1-phosphate and starch. For many years, it was assumed that this enzyme was involved in the synthesis of the α-(1 → 4)-linkages in starch, until Leloir and his co-workers (1961) first described starch synthetase. This latter enzyme catalyses the irreversible transfer of glucose from a nucleoside-diphosphate-glucose donor (NuDPG) to an existing chain of α-(1 → 4)-linked glucose residues where either ADPG or UDPG may function as donors:

$$NuDPG + [G]_n \rightarrow NuDP + [G]_{n+1}$$

Hence starch synthetase provides a chain-lengthening system which, together with Q-enzyme (branching enzyme), can synthesize starch. This discovery has necessitated a revision of the possible role of phosphorylase in starch metabolism. It certainly seems possible that in leaf tissues, which have a transient starch content, phosphorylase could efficiently convert starch to the hexose phosphate level, and hence to sucrose, for subsequent translocation. In storage tissues, the role of phosphorylase is not so clear-cut, since varying amounts of amylolytic enzymes are also present. Several workers, particularly Badenhuizen (1969), have been reluctant to ascribe a purely degradative role to phosphorylase and there have been suggestions that it is involved in the synthesis of primers for starch synthetase, since the enzyme cannot use glucose or maltose as acceptor substrates.

Phosphorylase had been isolated from many plant tissues and purified to varying extents, including isolation in a crystalline form (Holló et al., 1971). However, it is only in the last few years that the possible existence of multiple forms has been recognized.

The existence of possible isoenzymes of potato phosphorylase was first reported by Gerbrandy and Verleur (1971) and, subsequently, three peaks of activity were eluted from a DEAE-cellulose column (Gerbrandy and Doorgeest, 1972). Preliminary evidence of a dimer ⇌ tetramer relationship between some of the fractions was obtained but further work is clearly required since some of the peak fractions were shown to be heterogeneous when examined by electrophoresis.

Zea mays (sweetcorn) appears to contain four forms of phosphorylase which differ in certain physical properties, in rate of development and in their location (Table II; Tsai and Nelson, 1968, 1969). Phosphorylase I is present during all the stages of endosperm development and seed germination which were investigated, whilst phosphorylase II appeared only at the stage of rapid starch synthesis and was absent during germination. These two enzymes differed in many respects (see Table II), including behaviour towards nucleotides and other potential enzyme activators and inhibitors; moreover, the relative amount of phosphorylase II was ten

TABLE II

Properties of the maize phosphorylases[a]

	Isoenzyme			
Property	I	II	III	IV
First detectable in development	6 days	12–16 days	8–12 days	?
Present during germination	Yes	No	No	Yes
pH optimum	5·8	5·9	6·5	6·3
K_m (glucose-1-phosphate; mM)	3·3	4·0	2·0	1·0
Primer requirement	Yes	No	No	No
Relative activity at 22 days	100	1000	500	20

[a] Data from Tsai and Nelson (1969).

times that of phosphorylase I at 22 days after pollination. Phosphorylase III is present in both the endosperm and embryos of developing seeds but its activity is not easily detected since it complexes with a non-dialysable heat-labile inhibitor. Activity in the endosperm cannot be detected 8 days after pollination but is measurable after 12 days. Phosphorylase II is present only in the embryo of both developing and germinating seeds and behaves similarly to phosphorylase I on DEAE-cellulose chromatography but differs in pH optimum and electrophoretic mobility. The role of the enzymes *in vivo* has not been fully established but since phosphorylase II and III are detected during periods of rapid starch synthesis, they may be involved in the formation of primers. Phosphorylases I and IV are present in different tissues and may have a degradative role *in vivo*.

The phosphorylases present in an isogenic line of sweetcorn (*su* variety) have been examined by Lee (1972), who found evidence of multiple forms by gel electrophoresis, under conditions in which artefact formation was unlikely. Further results suggested that these bands arose from monomer-dimer forms which arose from three gene products. If *a*, *b* and *c* represent the three monomers, the endosperm contained cc, bc and bb whilst the embryo contained bb, ba and aa. Further developments in this study will be awaited with interest.

Phosphorylase has also been prepared from the leaves of mistletoe (Khanna *et al.*, 1971) and shown to exist in two forms A and B which differed in mobility on DEAE-cellulose, K_m values for starch and glucose-1-phosphate and behaviour towards AMP and phenolic inhibitors. It should be noted that enzyme action in normal leaf extracts was inhibited by endogenous phenolics and that special techniques were required to overcome this difficulty.

The presence of multiple forms of phosphorylase in various blue-green and green algae has been reported by Fredrick (1971). Species examined include *Oscillatoria princeps*, *Chlorella pyrenoidosa* and *Cyanidium caldarium*. However, the emphasis on this work has been on the biosynthesis of starch rather than on its degradation and is not therefore relevant to the present discussion. The examples are, however, of interest since it now seems that phosphorylases, whether from tubers, cereals, leaf tissues or algae, possess this general feature of existing in more than one form.

F. DEBRANCHING ENZYMES

None of the enzymes described so far has any action on the α-$(1 \rightarrow 6)$-glucosidic inter-chain linkages in amylopectin or α-dextrins. These linkages are hydrolysed by specific α-$(1 \rightarrow 6)$-glucosidases which are collectively termed debranching enzymes. With amylopectin, this hydrolysis results in limited increases in iodine staining power and β-amylolysis limit and a relatively small increase in reducing power which may be at or below the sensitivity of the analytical reagent (Dunn *et al.*, 1973). With α-dextrins, the increase in reducing power is greater and can be measured more easily. Some plant debranching enzymes which readily hydrolyse α-dextrins also act on pullulan (a fungal polysaccharide containing α-$(1 \rightarrow 6)$-linked maltotriose residues) so that this substrate can be used to assay the enzymes (Manners *et al.*, 1970).

The subject of the specificity and nomenclature of debranching enzymes has been the subject of some controversy, but it is not proposed to discuss these aspects in detail in this review. In general terms, some debranching enzymes readily attack amylopectin and its β-limit dextrin but have little or no action on α-dextrins or pullulan. This activity will be referred to as amylopectin 6-glucanohydrolase, in preference to the former term, R-enzyme, which has been redefined since the original report of Hobson *et al.* (1951). Other debranching enzymes have a high affinity towards α-dextrins and pullulan, but show only limited activity towards amylopectin; these enzymes will be described as limit dextrinases, a term which has been used in the literature for almost 30 years. An alternative suggestion (Lee and Whelan, 1971) to describe this activity as plant pullulanase can only lead to confusion.

In some plant extracts, more than one debranching enzyme is present but since these have different specificities, they cannot be regarded as isoenzymes. Typical results for enzyme preparations from broad beans and malted barley are shown in Fig. 3. Enzyme activity which increased the iodine staining power of amylopectin (i.e. amylopectin 6-glucanohydrolase or R-enzyme) is separated from that towards α-limit dextrins (MacWilliam and Harris, 1959). It should be noted that the fractions showing limit dextrinase activity also show amylase activity, so that this would interfere with attempts to determine whether or not the limit dextrinase could act on amylopectin. The original statement of these authors that limit dextrinase "appears to be without action on larger substrates" is therefore incorrect. Nevertheless, those fractions which showed amylopectin 6-glucanohydrolase activity did not act on α-limit dextrins, so that their conclusion that two debranching enzymes were present is perfectly justified.

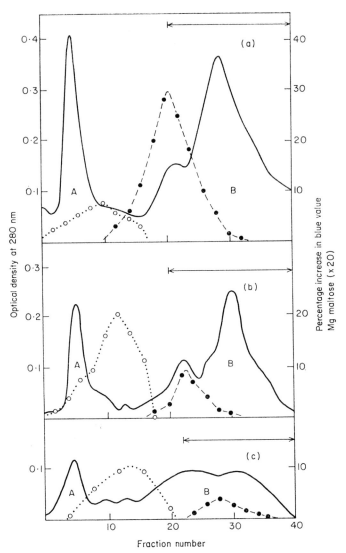

FIG. 3 Distribution of protein and enzymic activities of (a) malt enzyme, (b) broad bean enzyme, and (c) R-enzyme concentrates eluted from alumina columns. Optical density at 280 nm as a measure of protein concn ——. R-enzyme activity as percentage increase in blue value after 6 h at 37°. ○ ···· ○ Limit dextrinase activity expressed as mg maltose/ml of enzyme fraction produced from α-limit dextrins. ● – – – ● Amylase activity as revealed by chromatographic examination ←—→. Reproduced with permission from MacWilliam and Harris (1959)

At the present time, the only evidence for the existence of isoenzyme forms of a debranching enzyme appears to be that of Lee *et al.* (1971) who showed that chromatography of a sweetcorn limit dextrinase preparation on hydroxylapatite gave two components. Both of these readily hydrolysed α-limit dextrins, amylopectin β-limit dextrin and pullulan but hydrolysed amylopectin at only about 6–8% of the rate of the α-limit dextrins. Sweetcorn appears to be unique amongst the higher plants in storing both a two-component starch and phyto-glycogen as carbohydrate reserves. Extracts of sweetcorn also contain two other debranching enzymes, an amylopectin 6-glucanohydrolase (which acts on amylopectin but not on phytoglycogen) and an isoamylase (which can hydro-lyse the inter-chain linkages in both polysaccharides; see Manners and Rowe, 1969). The function *in vivo* of these two enzymes and their relationship to the limit dextrinase isoenzymes is not yet known.

G. CONCLUSIONS

During the last decade, which has seen spectacular progress in our under-standing of the complex systems for the biosynthesis of proteins and nucleic acids, there has been an impression that knowledge of the metabolism of polysaccharides, and particularly of starch, was essentially complete and that the metabolic pathways which have appeared in many textbooks and review articles accurately described the situation *in vivo*. This is clearly not the case. There is now substantial evidence that many of the starch-degrading enzymes exist in more than one form, but questions as to the site of biosynthesis and precise function *in vivo* of the individual forms remains for future experimental investigations. It would seem wasteful for nature to provide duplicate enzymes for a single pathway, unless they served as part of a sophisticated regulatory control system. In many cases, there is no evidence to suggest that the multiple forms represent active and inactive forms of the same enzyme, the interconver-sion of which could provide a means of regulating a metabolic pathway, as for example with phosphorylase and glycogen synthetase in glycogen metabolism.

Future work will require careful experimentation to locate the site of biosynthesis of the individual enzymes in the plant tissues, and their site(s) of action. The use of flour from whole cereal grains may provide information on the overall situation but is clearly unsuitable for *in vivo* studies. An interest-ing example of the type of work which is required has been reported by MacGregor *et al.* (1972) who carried out dissection studies on two cultivars of barley and showed that the pericarp of developing barley kernels contained both α-amylase and starch granules. Soon after anthesis the starch completely disappeared and the α-amylase activity then decreased rapidly to a very low level which was that of the mature kernels. This enzyme is, therefore, physio-logically different from the major α-amylase of germinated barley which arises from *de novo* synthesis in the aleurone layer. Hence it is probable that the function of the pericarp enzyme is to provide energy for the growing kernel by hydrolysing the pericarp starch.

Similar observations on wheat have been described by Banks *et al.* (1972), who have observed that the α-amylase present in the initial growth stage (8–14 days after anthesis) is almost entirely in the pericarp–testa fraction of the grain and that the amount in the endosperm is negligible. Further studies of this type, particularly with respect to phosphorylase, β-amylase and the debranching enzymes, are clearly required.

III. Starch Degradation During the Germination of Cereals

During the germination of cereals, a number of significant changes in the starch take place. These include a decrease in the starch content of the kernel, an apparent increase in the amylose content of the starch, and limited decreases in the molecular size of the amylose and in the average chain length and β-amylolysis limit of the amylopectin. On the basis of the analytical changes in barley starch, Greenwood and Thomson (1959, 1961) suggested that these changes arose from a limited β-amylolysis of the amylopectin component and a limited α-amylolysis of the amylose. Manners and Bathgate (1969) considered this explanation to be unlikely in view of the fact that β-amylase, acting alone, has little or no action on intact starch granules; the explanation also implies some inherent selectivity in the action of one type of amylase on only one of the two starch components. These workers carried out an analogous study on oat starch, and found the same changes in the properties of the two components as had been previously observed with barley starch (see Table III). However, it was suggested that the above changes could be brought about by limited α-amylolysis of both components and this view was supported by subjecting oat starch granules to limited α-amylolysis; the changes in the properties of the amylose and amylopectin components were similar to those observed during malting. It was, therefore, concluded that α-amylase was the predominant enzyme involved during the germination process.

Table III

Properties of starches from ungerminated and malted barley and oats[a]

Property	Barley	Malted barley	Oats	Malted oats
WHOLE STARCH				
Starch content of kernel ($\%$ dry weight)	64	58	52	48
Iodine affinity ($\%$)	4·25	4·95	4·1	5·1
Amylose content ($\%$)	22	26	22	28
AMYLOSE COMPONENT				
Iodine affinity ($\%$)	19·0	19·0	18·5	18·4
Limiting viscosity number	240	115–200	345	256
Degree of polymerization	1800	850–1480	2550	1900
β-Amylolysis limit ($\%$)	72	77–90	82	90
AMYLOPECTIN COMPONENT				
Iodine affinity ($\%$)	0·4	0·7	0·5	0·4
Average chain length	26	18	20	18
β-Amylolysis limit ($\%$)	58	48	56	49

[a] Data for barley from Greenwood and Thomson (1959) and for oats from Manners and Bathgate (1969).

The situation has recently been further clarified by research workers at the Brewing Industry Research Foundation. Palmer (1972) confirmed earlier reports that barley starch contains two morphologically distinct types of granule, i.e. large round granules with an average diameter of about 25 μm and much smaller granules of about 5 μm in diameter. During malting, the amylases have only a limited action on the large granules but the small granules are rapidly degraded. These small granules represent about 90% of the total starch granules *by number*, but only about 10% *by weight*. The two types of granule were separated by Bathgate and Palmer (1972) and characterized. The results (Table IV) show that the starch from the small granules differs from that of the larger granules. The previous results of Greenwood and Thomson (1959, 1961) represent, in fact, the properties of the starch from the larger granules since the smaller granules were inadvertently discarded during the isolation and purification of the starch. By

TABLE IV

Properties of large and small starch granules from barley and wheat[a]

Starch sample	Iodine affinity (%)	Amylose content (%)	β-Amylolysis limit (%)
BARLEY			
Large granules	4·98	25	64
Small granules	8·25	41	80
WHEAT			
Large granules	5·01	25	72
Small granules	4·79	24	—

[a] Data from Bathgate and Palmer (1972).

scanning electron microscopy, it has been shown that there is limited α-amylolytic attack on the large granules (Palmer, 1972) as shown by the appearance of radial channels. This action could account for the observed changes in the structure of the amylopectin component and also of the amylose, part of which may be in close proximity to the branched components.

Although most samples of barley starch contain about 26% of amylose, some varieties, e.g. Glacier, produce a starch with a much higher amylose content of about 47%. It is of interest to note that in the high amylose barley, the small starch granules are absent and the population and size of the larger granules is less than in normal barley starch (Palmer, 1972).

In wheat starch, there is a range of granule size but no bimodal distribution similar to that observed in barley (Bathgate and Palmer, 1972). In this cereal, the properties of the large and small granules are generally similar (see Table IV) and during malting there is no preferential attack on the small starch granules. This difference may be due, in part, to the close association of the small wheat starch granules with protein material.

The overall results show unexpected complexities in the pathways for the synthesis and degradation of starch in some cereals. It is possible, for example, that in barley the synthesis of the two sizes of granule may be under separate genetic control. The results provide further evidence for the view that future progress must depend on a careful examination of the distinct parts of the cereal grain and that the use of cereal flours as experimental material can only give limited information.

ACKNOWLEDGEMENT

This review was prepared whilst the writer was on sabbatical leave in the Department of Biochemistry, University of Leicester, and the kind hospitality of Professor Hans Kornberg, F.R.S. is gratefully acknowledged.

REFERENCES

Badenhuizen, N. P. (1969). "The Biogenesis of Starch Granules in Higher Plants." Appleton-Century-Crofts, New York.
Banks, W., Greenwood, C. T. and Khan, K. M. (1971). *Carbohyd. Res.* **20**, 233.
Banks, W., Evers, A. D. and Muir, D. D. (1972). *Chemy Ind.* 573.
Bathgate, G. N. and Palmer, G. H. (1972). *Die Stärke* **10**, 336.
Bilderback, D. E. (1971). *Pl. Physiol.* **48**, 331.
Botes, D. P., Joubert, F. J. and Novellie, L. (1967a). *J. Sci. Fd Agric.* **18**, 415.
Botes, D. P., Joubert, F. J. and Novellie, L. (1967b). *J. Sci. Fd Agric.* **18**, 409.
Chao, S. E. and Scandalios, J. G. (1972). *Molec. gen. Genetics* **115**, 1.
Dunn, G., Hardie, D. G. and Manners, D. J. (1973). *Biochem. J.* **133**, 413.
Fekete, M. A. R. de (1968). *Planta* **79**, 208.
Fredrick, J. F. (1971). *Phytochemistry* **10**, 395.
Frydenberg, O. and Nielsen, G. (1965). *Hereditas* **54**, 123.
Gerbrandy, S. J. and Doorgeest, A. (1972). *Phytochemistry* **11**, 2403.
Gerbrandy, S. J. and Verleur, J. D. (1971). *Phytochemistry* **10**, 261.
Greenwood, C. T. and Milne, E. A. (1968). *Adv. Carbohyd. Chem.* **23**, 281.
Greenwood, C. T. and Thomson, J. (1959). *J. Inst. Brew.* **65**, 346.
Greenwood, C. T. and Thomson, J. (1961). *J. Inst. Brew.* **67**, 64.
Hanes, C. S. (1940). *Proc. R. Soc.* B **128**, 421.
Hobson, P. N., Whelan, W. J. and Peat, S. (1951). *J. chem. Soc.* 1451.
Holló, J., László, E. and Hoschke, A. (1971). "Plant α-(1 → 4)-Glucan Phosphorylase." Akademiai Kiado, Budapest.
Hutson, D. and Manners, D. J. (1965). *Biochem. J.* **94**, 783.
Jacobsen, J. V., Scandalios, J. G. and Varner, J. E. (1970). *Pl. Physiol.* **45**, 367.
Jørgensen, O. B. (1963). *Acta chem. scand.* **17**, 2471.
Jørgensen, O. B. (1964). *Acta chem. scand.* **18**, 1975.
Jørgensen, B. B. and Jørgensen, O. B. (1963). *Acta chem. scand.* **17**, 1765.
Keller, P. J., Kauffman, D. L., Allan, B. J. and Williams, B. L. (1971). *Biochemistry* **10**, 4867.

Khanna, S. K., Krishnan, P. S. and Sanwal, G. G. (1971). *Phytochemistry* **10**, 545, 551.
Kruger, J. E. (1970). *Cereal Chem.* **47**, 79.
Kruger, J. E. and Tkachuk, R. (1969). *Cereal Chem.* **46**, 219.
LaBerge, D. E. and Meredith, W. O. S. (1969). *J. Inst. Brew.* **75**, 19.
LaBerge, D. E. and Meredith, W. O. S. (1971). *J. Inst. Brew.* **77**, 436.
Lee, E. Y. C. (1972). *FEBS Letters* **27**, 341.
Lee, E. Y. C., Marshall, J. J. and Whelan, W. J. (1971). *Archs Biochem. Biophys.* **143**, 365.
Lee, E. Y. C. and Whelan, W. J. (1971). *In* "The Enzymes", 3rd edition (P. D. Boyer, ed.), Vol. 5, p. 191. Academic Press, New York and London.
Leloir, L. F., de Fekete, M. A. R. and Cardini, C. E. (1961). *J. biol. Chem.* **236**, 636.
MacGregor, A. W., Gordon, A. G., Meredith, W. O. S. and Lacroix, L. (1972). *J. Inst. Brew.* **78**, 174.
MacGregor, A. W., LaBerge, D. E. and Meredith, W. O. S. (1971). *Cereal Chem.* **48**, 490.
MacGregor, A. W. and Meredith, W. O. S. (1971). *J. Inst. Brew.* **77**, 510.
MacWilliam, I. C. and Harris, G. (1959). *Archs Biochem. Biophys.* **84**, 442.
Manners, D. J. (1962). *Adv. Carbohyd. Chem.* **17**, 371.
Manners, D. J. and Bathgate, G. N. (1969). *J. Inst. Brew.* **75**, 169.
Manners, D. J. and Marshall, J. J. (1972). *Die Stärke* **24**, 3.
Manners, D. J., Marshall, J. J. and Yellowlees, D. (1970). *Biochem. J.* **116**, 539.
Manners, D. J. and Rowe, K. L. (1969). *Carbohyd. Res.* **9**, 107.
Markert, C. L. and Møller, F. (1959). *Proc. natn. Acad. Sci. Wash.* **45**, 753.
Marshall, J. J. and Taylor, P. M. (1971). *Biochem. biophys. Res. Commun.* **42**, 173.
Palmer, G. H. (1972). *J. Inst. Brew.* **78**, 326.
Payne, D. M., Porter, D. W. and Gracy, R. W. (1972). *Archs Biochem. Biophys.* **122**, 122.
Robyt, J. F. and Whelan, W. J. (1968). *In* "Starch and Its Derivatives", 4th edition (J. A. Radley, ed.), p. 423. Chapman and Hall, London.
Sanders, T. G. and Rutter, W. J. (1972). *Biochemistry* **11**, 130.
Siepmann, R. and Stegemann, H. (1967). *Z. Naturf.* **22**, 949.
Takahashi, N., Shimomura, T. and Chiba, S. (1971). *Agric. biol. Chem.* **35**, 2015.
Tsai, C. Y. and Nelson, O. E. (1968). *Pl. Physiol.* **43**, 103.
Tsai, C. Y. and Nelson, O. E. (1969). *Pl. Physiol.* **44**, 159.

CHAPTER 9

Starch Metabolism: Synthesis versus Degradation Pathways

MARIA A. R. DE FEKETE AND GEORG H. VIEWEG

Technische Hochschule Darmstadt, Fachbereich Biologie
Darmstadt, West Germany

I. INTRODUCTION AND HISTORY

The role of enzymes involved in starch metabolism has always been a source of speculation and hypothesis. For many years, it was assumed that polysaccharides were formed by reversal of hydrolysis and in order to explain how such syntheses could operate, it was necessary to suppose that in certain zones of cells the concentration of monosaccharides was very high.

With the discovery of phosphorylase a new controversy began. The reaction catalysed by this enzyme is reversible and, hence, both formation and breakdown of glucans are possible. To synthesize, or to degrade: that was the question: why would the reaction go in one direction or in the other? However, this was an indiscreet question. Investigators should never ask why, if they want to obtain an answer!

In the late fifties the hopes for a safe world, with well defined metabolic pathways for starch synthesis, were centred on glucosyl transfer from sugar nucleotides. The enzyme catalysing this reaction turned out to be very elusive. In quite a number of early, frustrating experiments with different plant materials the incorporation of glucose from UDPG into starch could not be demonstrated. We now know that the difficulties encountered in detecting synthetase activity were due to the use of UDPG as a glucosyl donor and because the removal of cell debris from the enzyme preparations also removed the starch granules which, it was eventually discovered, contained the starch-synthesizing enzyme. The discovery of the active synthetase cast doubts on the synthetic role of phosphorylase (Fekete *et al.*, 1960): the apparent ratio of inorganic phosphate to glucose-1-phosphate in the cell which was known to be unfavourable for synthesis (Ewart *et al.*, 1954), and the fact that Stocking (1952) could not detect phosphorylase in chloroplasts were both remembered. Even arguments concerning animal tissues were used to support the theory that phosphorylase was only involved in starch breakdown. In the excitement of the discovery of synthetase we debated the role of the enzymes involved in starch metabolism only in terms of cells consisting of a simple mixture of enzymes, substrates and products. Biochemists often make similar simplifications based on the reactions of purified enzymes *in vitro* under well defined conditions which are not necessarily those which exist in the cell at the site where the enzyme is located. From the results of such ideal experiments inaccurate extrapolations are made to metabolic pathways *in vivo*. The biologists are also not without fault. Their determinations of enzymic activities and concentrations of substrates and products under different conditions or in diverse tissues furnish only a very restricted view of most processes.

In the years that have elapsed since the discovery of starch synthetase our ideas have slowly changed. It is now recognized that redundant pathways and multiple interactions at each metabolic step exist in cells. Starch metabolism is something more than just synthesis and degradation reactions. An integration of biochemical and botanical ideas is unavoidable. At the present time we should, therefore, try to examine the enzymes without forgetting that they are constituents of parts of cells which in turn form parts of the plant.

II. Interlude and Present Knowledge

The similarities between starch and animal glycogen must have induced quite a number of investigators to search for a parallelism between the metabolism of these polysaccharides and many findings encouraged such comparisons. Thus, amylases, phosphorylase and synthetase were found in both animal and plant tissues. But when attempts were made to understand the details of the metabolic pathways all these generalizations became a handicap.

The requisites for the life of plants and animals are not comparable. As stated by Smith *et al.* (1968), animal glycogen is an energy reserve capable of rapid mobilization. Because of its highly branched structure, it is very soluble and, therefore, easily accessible to degrading enzymes. Moreover, as a result of the extensive branching in the glycogen molecule about 10% of all glucose units are available for immediate release as glucose-1-phosphate. On the contrary, the metabolic turnover of starch in plants is a slow process

(Badenhuizen, 1971) and only under favourable conditions in rapidly metabolizing tissues can it be measured in terms of hours. Starch formation or degradation extending over days or weeks is no exception. This process takes place in the plastids where certain metabolites may be concentrated while others are excluded. Porter (1962) pointed out that the ratio of inorganic phosphate to glucose-1-phosphate measured in tissues may not correspond to the value for plastids. In plant cells the turnover of starch is often coupled with that of sucrose, the transport sugar of plants. This is probably a further factor which distinguishes between the metabolism of starch and glycogen.

At the beginning of our investigation of the metabolism of starch we asked ourselves what we really knew about the function of amylases, phosphorylase and synthetase? The facts were:

(1) Amylases catalyse the breakdown of starch.
(2) At least some of the starch is formed from sugar nucleotides.
(3) Phosphorylase could be involved in synthesis and breakdown.

A possible approach to the problem is to attempt to correlate enzymic activities of cells metabolizing starch with changes in starch content *in vivo* under defined conditions. One could expect that synthesis would be accompanied by an increase in the levels of enzymes catalyzing starch formation and that during breakdown the activity of degrading enzymes would rise. At first sight it would seem that such experiments are very easy to perform. There are indeed quite a number of papers on such observations, but a careful examination of this published data often shows that some important information is lacking, perhaps due to the fact that the authors had other aims in mind whilst performing their experiments. For instance, in our first papers dealing with the synthesis of starch in bundle sheath cells of maize leaves, we were concerned with the possible involvement of phosphorylase in starch synthesis and were not interested in the very active amylases in these cells. We therefore investigated the activities of starch synthetase and phosphorylase and found that the former was too low to account for the starch formed *in vivo* (Fekete and Vieweg, 1971). Low synthetase activities were also observed in developing *Vicia faba* cotyledons (Fekete, 1969a), in chloroplasts (Doi and Doi, 1969) and in potato tubers (Potter, 1964). Failure to detect enzymes or finding low activities is always disturbing: such results can be a consequence of poor assay techniques. Doubts can be partially allayed by repeated confirmation of the negative findings by different workers using various materials and different assay procedures. More reliance can be given to negative results when they can be compared with positive observations. Thus Badenhuizen and Chandorkar (1965) showed that potato tubers grown at low temperatures possessed synthetase activity, but culture at 30°C brought about a disappearance of the enzyme although the production of starch was normal.

We believe that all generalizations should be avoided, even if the temptation is great. It would be erroneous, for example, if we were to conclude that the presence of synthetase is not important for starch synthesis. There are many tissues where this enzyme is so active that there is no need to postulate the existence of others in order to account for the amount of starch that accumulates *in vivo*: such is the case with developing rice grains (Baun *et al.*, 1970), maize endosperm (Ozbun *et al.*, 1973) and nectaries (Vieweg, unpublished results). High phosphorylase activity is characteristic not only of tissues where starch synthetase activity is low but also of those where a rapid glucosyl transfer from sugar nucleotides can be demonstrated. It is inter-

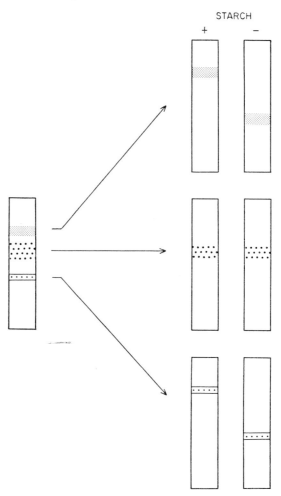

FIG. 1. Gel-electrophoresis of phosphorylases from immature *Vicia faba* cotyledons. Left: Crude extract. Right: Second gel electrophoresis, in presence and absence of starch, of the three bands from the crude extract isolated by extraction and precipitation with ammonium sulphate (data taken from Rebenich, 1972).

esting that in most tissues more than one phosphorylase can be detected but we found that it was difficult to define the exact number: to assume that more than two are present is always hazardous. Studies in our laboratory have shown that these fears are well founded. Rebenich (1972) has demonstrated that one of the phosphorylases in *Vicia faba* cotyledons can bind tightly to different primers. These enzyme–primer complexes are very stable and they do not dissociate during some enzyme purification procedures; they can be observed as distinct bands on polyacrylamide gel disc electropherograms: Fig. 1 shows such an experiment. Working with a crude enzyme preparation from developing *Vicia faba* cotyledons, three phosphorylase bands could be detected in a first electrophoretic separation. The separated enzymes were extracted from the gels, precipitated with ammonium sulphate and each fraction again examined by gel electrophoresis in the presence or absence of starch. After this treatment only the middle band from the first separation maintained its original electrophoretic mobility. The migrations of the original fast and slow bands in the second separation were dependent on the presence or absence of starch. Both showed the same fast migrating band in the absence of starch, whilst in its presence, the migration corresponded to the original slow-moving component. A careful revision of the already reported data on the multiplicity of phosphorylases should, therefore, be carried out considering the possibility that some of the observed differences in the properties of the enzymes may be due to their association with different primers and that these in turn depend on the growth conditions or developmental state of the tissue. The two phosphorylases, I and II, of *Vicia faba* cotyledons can be separated from each other by repeated precipitation with ammonium sulphate (Fekete, 1968). Their behaviour towards various primers shows up some differences. In Table I it can be seen that only phosphorylase II is able to catalyse glucosyl transfer from glucose-1-phosphate to starch granules. This phosphorylase has a lower K_m for amylose and amylopectin

TABLE I

Effect of different primers on the activity of the phosphorylases of *Vicia faba*[a]

	Glucose incorporation (cpm)	
Primer	Phosphorylase I	Phosphorylase II (starch granule phosphorylase)
Starch granules	130	2223
Amylopectin	6185	2601
Amylose	1477	3952

[a] The experiment was performed as explained in the legends for Figs 2 and 3, but using starch granules (15 mg), amylopectin (7 μg) or amylose (25 μg).

than phosphorylase I and is inhibited at high amylopectin concentrations (phosphorylase I is not inhibited by amylopectin).

The effect of amylose and amylopectin concentration on the glucose incorporation from glucose-1-phosphate catalysed by both phosphorylases can be observed in Figs 2 and 3. The most striking property of phosphorylase I is the high activity with maltooligosaccharides as primers.

In presence of some salts (NaCl, KCl,(NH$_4$)$_2$SO$_4$, EDTA-Na) phosphorylase II can be adsorbed on starch granules or even chloroplasts (Fekete, 1968). Scarcely any adsorption takes place if the salt medium is replaced by a

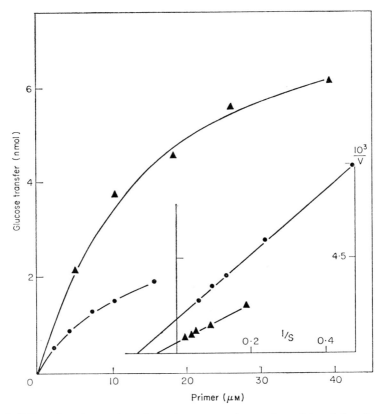

FIG. 2. Effect of amylose and amylopectin concentration on the activity of phosphorylase I from *Vicia faba* cotyledons. Phosphorylase I was separated from extracts of immature *Vicia faba* cotyledons as described by Fekete (1968) and purified about 50-fold by repeated dilution and reprecipitation with ammonium sulphate. The incubation mixture contained: citrate buffer (pH 6·1; 5 μmol); ^{14}C-glucose-1-phosphate (26 500 cpm, 0·12 μmol); primer and enzyme (23 μg protein) in a total volume of 50 μl. After 5, 10 and 15 min the incubation was stopped with 75% methanol. Subsequent further treatment and calculations were as described by Fekete and Cardini (1964). The concentration of the primers are expressed as non-reducing end groups assuming 4·5% for amylopectin (▲) and 0·32% for amylose (●). The K_m value derived from the graph for amylose is 9·1 μM and for amylopectin, 16 μM. V_{max} amylopectin/V_{max} amylose = 3.

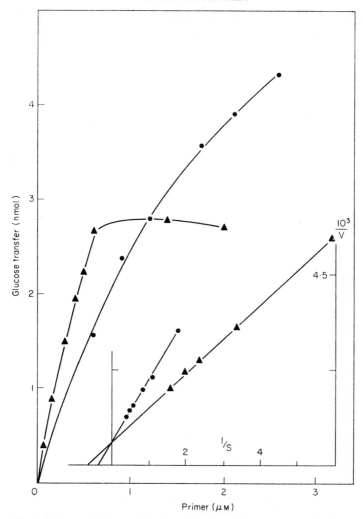

FIG. 3. Effect of amylose and amylopectin concentration on the activity of phosphorylase II from *Vicia faba* cotyledons. The experimental conditions and symbols are as described in Fig. 2. Assays were performed with 40 μg protein. The following values were derived from the plot: K_m for amylose, 3·2 μM, and for amylopectin, 1·5 μM; V_{max} amylopectin/ V_{max} amylose = 1.

sucrose medium during the adsorption process, although sucrose has no influence on the activity of the already adsorbed enzyme (Fekete, 1966). We suggest that the trivial name, "starch granule phosphorylase", should be used for phosphorylase II.

One may ask if the starch produced by adsorbed phosphorylase is in the same location in the granule and corresponds in structure to the polysaccharide formed by the starch granule-bound synthetase. Table II shows that this is not the case. The radioactivity introduced into the starch from ^{14}C-glucose-1-phosphate by phosphorylase is readily solubilized by the same enzyme. However, the starch formed by synthetase from ADP-^{14}C-glucose is less readily phosphorylysed (Fekete, 1969a). These results can be related to experiments by Gafin and Badenhuizen (1959). Using ^{14}CO$_2$ they labelled starch granules from tobacco leaves and showed that most of the radioactivity incorporated in the light was lost in a subsequent dark period. The authors suggested that only a small amount of ^{14}C penetrated beyond the outer layers of the granules and was not accessible to the action of reagents and enzymes. We suppose that this inert radioactive starch could be formed from ADP-glucose by synthetase, which is evenly distributed throughout the granule (Badenhuizen, 1965), whilst that which can be easily removed might arise from the activity of phosphorylase. All these results suggest that the participation of both phosphorylase and synthetase in starch synthesis are possible and that both enzymes may even function simultaneously.

TABLE II

Action of *Vicia faba* cotyledon preparations containing phosphorylase II on starch granules labelled from ^{14}C-glucose-1-phosphate and ADP-^{14}C-glucose[a]

Source of starch granule label	Incubation buffer	Solubilization by phosphorylase II %
^{14}C-glucose-1-P	Phosphate	61·8
^{14}C-glucose-1-P	Citrate	45·0
ADP-^{14}C-glucose	Phosphate	21·4
ADP-^{14}C-glucose	Citrate	21·8

[a] Data taken from Fekete, 1969a.

Whilst observing the enzymic patterns of starch-synthesizing tissues, we often noted high amylase activities. One would expect amylase activity to be high during starch breakdown but sometimes this activity was so high in extracts of starch-synthesizing tissues as to suggest no starch could accumulate if these amylases actually functioned at the site of synthesis. We have already mentioned the amylase activity in the bundle sheath cells of maize and there are quite a number of other publications that report high amylase activity in starch-synthesizing tissues (Baun *et al.*, 1970; Haapala, 1969; Gates and Simpson, 1968).

Another astonishing finding was the presence of synthetase in cotyledons of germinating peas (Chandorkar and Badenhuizen, 1967) and broad beans (Fekete, 1969b). Juliano and Varner (1969) found an incorporation of radioactivity from glucose-^{14}C into the starch of the cotyledons of germinating peas. This indicates that starch is synthesized even during the rapid phase of starch degradation. On the other hand, degradation accompanies synthesis.

For example, Chan and Bird (1960) found that "when tobacco leaves containing radioactive starch were kept in air and at light intensity which maintained a constant amount of starch, there was an exchange of radioactive carbon from the starch". All these results indicate that the metabolic pathways are far more complicated than the overall balance of metabolites or variations in the enzymic pattern, reflect. Starch metabolism is the sum of quite a number of different synthetic and degradative steps which all function at approximately the same time. It is obvious that only an integration of all the known facts would allow us to gain an insight into the mystery of starch.

III. THEME AND VARIATIONS

In order to solve a mystery detectives can gather facts and testimonies and with a great deal of patience and luck, these can often be pieced together and a suspect arraigned. Another method is to induce the suspect to act in such a way that self-incrimination results. Both approaches can be made to the starch mystery. Information concerning amylose and amylopectin can be gathered and all enzymes that may possibly be involved in the metabolism of these polysaccharides can be examined to see how the presence of metabolites changes their activities. It is also possible to envisage experiments where the production or degradation of starch is achieved in a system simulating conditions *in vivo* or by bringing about a reversal of the metabolism of cells. The information gained from such investigations leads to further understanding of the pathways only if it is in an agreement with the knowledge that has accumulated from other approaches to the problem. In this case it may show up details of interactions and control mechanisms that could not otherwise have been observed.

Reserve tissues, such as cereal grain endosperms, leguminous seed cotyledons and potato tubers, synthesize starch from sucrose which is transported from the photosynthesizing leaves; in a later physiological stage starch is degraded and sucrose produced. There are quite a number of publications that describe how such sucrose–starch and starch–sucrose transformations may proceed (Fekete and Cardini, 1964; Murata *et al.*, 1964; Fekete, 1969a,b; Shannon, 1968; Viswanathan and Krishnan, 1966; Turner, 1969). The problem is that all of these transformations proceed slowly and thus the control mechanisms are limited. In the hope of finding a new approach to this metabolic puzzle we started our experiments with leaves that easily "switch" from starch synthesis to degradation and vice versa.

For this study we chose maize leaves which have some characteristic features relating anatomy, morphology, physiology and CO_2-fixation patterns. In these leaves two types of photosynthetic cells may be found: the mesophyll cells with chloroplasts that under normal conditions are nearly devoid of starch, and the bundle sheath cells with big plastids lacking grana and photosystem II that on illumination produce starch (cf. Hatch *et al.*, 1971). The bundle sheath cells can be easily separated from the surrounding mesophyll and once isolated they are no longer able to produce starch unless the suspending medium is supplied with glucose-1-phosphate. The presence of sucrose, glucose, maltose or sugar phosphates does not, either in the dark or in the light, lead to starch accumulation (Fekete and Vieweg, 1971; Vieweg and Fekete, 1972). In leaf strips where the bundle sheath cells are still in contact with the mesophyll, starch can be produced from different sugars but only in the light (Vieweg and Fekete, 1973). These results would suggest that bundle sheath cells behave as reserve tissues and that they depend on the transport of a metabolite from the mesophyll in order to produce starch. The main translocatory sugar is sucrose (Ziegler, 1956) and the transport of this disaccharide is advantageous for the cells not only from the energetic point of view but also because it helps with the maintainance of the ionic equilibrium. In the case of bundle sheath cells, the transported metabolite is glucose-1-phosphate. As a consequence of this transport new problems had to be solved and in particular the question of what happens to the phosphate liberated in the course of starch synthesis. Our experiments showed that when isolated bundle sheath cells are suspended in a solution of glucose-1-phosphate starch is formed and the amount of glucose-1-phosphate that disappears from the solution is equivalent to the amount of inorganic phosphate appearing in the medium (Fig. 4). The results also indicated that immediately after starch is formed, some of it is hydrolysed, giving rise to maltose (also some glucose, maltotriose and other oligosaccharides) which appears in the medium. We already knew that bundle sheath cells contained active amylases, so we were not surprised by this degradation process which accompanied starch synthesis.

In further experiments we showed that the accumulation of starch from glucose-1-phosphate in isolated bundle sheath cells was increased by the addition of maltose to the incubation mixture (Vieweg and Fekete, 1973; Fekete and Vieweg, 1973b). Using radioactive maltose we did not find any incorporation of label into the starch or oligosaccharides formed and there was little evidence to suggest that the disaccharide was metabolized at all. These results suggested that maltose was inhibiting the activity of amylase and this supposition was confirmed using enzymic preparations from isolated bundle sheath cells. Schwimmer (1950) had previously reported that malt α-amylase was inhibited by maltose; also, the activity of sweet potato β-amylase is decreased by the presence of 0·15 M maltose (French and Youngquist, 1963). When isolated bundle sheath cells were incubated with ^{14}C-glucose-1-phosphate the production of radioactive maltose, maltotriose and glucose decreased when maltose was added to the medium. These latter results also strengthen the assumption that maltose is controlling the activity of the amylase.

We were more impressed by the results of experiments where we tried to follow the breakdown of the starch that plants had accumulated in the light.

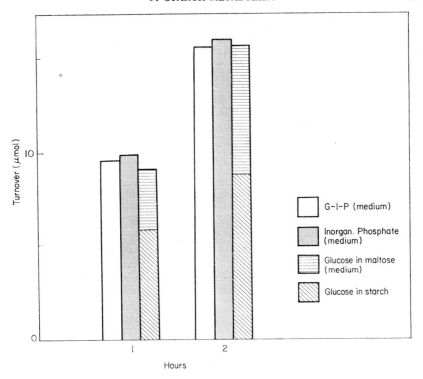

FIG. 4. Stoichiometry of the starch accumulation process in isolated bundle sheath cells from maize leaves. The experiment was performed as described by Vieweg and Fekete (1973). The cells were suspended in 10 mM glucose-1-phosphate solution at pH 5·7.

Leaves attached to the plants lose most of their starch in one night. If, however, the leaves were detached after the illumination period and the cut ends dipped in water, then only half as much starch disappeared. Further, when the leaves were cut into strips 0·5 mm wide and left in water in the dark, the starch content scarcely changed even after a 16–20 h incubation nor did starch disappear from isolated bundle sheath cells when these were suspended in water. For some time we could not find an explanation for these results that seemed to be contradictory. On the one hand, starch production from glucose-1-phosphate in isolated bundle sheath cells was accompanied by a simultaneous breakdown and on the other, once starch had accumulated in the whole leaf, then the isolated bundle sheath cells would no longer degrade it.

This suggested to us that the amylase activity might disappear in the light because in order to follow starch formation we used bundle sheath cells prepared from leaves of plants that had been kept in the dark for 2 days and when investigating starch breakdown we used material that had been illuminated for 24 h. However, we were disappointed when we observed that amylases were present in both bundle sheath preparations coming from light- and from dark-treated plants. In a series of experiments we tried by all the means that were available to induce the bundle sheath cells to degrade their own starch. In the Table III we have compiled the most interesting data obtained from these experiments.

In order to detect a breakdown of starch in the isolated bundle sheath cells the pH value of the suspending medium should be above 7. Table III shows that the addition of some form of phosphate enhances starch degradation. This phosphate may be fructose-1,6-diphosphate or inorganic phosphate which was only effective at certain concentrations (about 10–20 mM). Unex-

TABLE III

Starch breakdown in maize leaves

Material	Addition (10 mM)	Starch breakdown (μmol glucose/g fresh wt tissue/h)
Experiment A. Plants illuminated for 34 h		
Leaf attached	—	1·9
Leaf detached	—	0·73
Bundle sheath[a]	none	0·34
Bundle sheath	P_i	0·84
Bundle sheath	Fru-1,6-diphosphate	1·22
Experiment B. Plants illuminated for 58 h		
Bundle sheath	none	1·28
Bundle sheath	P_i	3·52
Bundle sheath	Fru-1,6-diphosphate	2·13
Bundle sheath	G-1-phosphate	2·58

[a] Bundle sheath cells were prepared and the starch content measured as previously described (Vieweg and Fekete, 1972). Each digest contained cell preparation (300 mg) in 40 mM morpholinopropane sulphonic acid buffer (3 ml, pH 7·5) and additions as indicated: incubations were carried out for 20 h at 30°. Starch content of bundle sheath cells: experiment A, 7·9; experiment B, 24·6, mg/g fresh wt tissue.

pectedly, the presence of 10 mM glucose-1-phosphate at pH 7·5 in the incubation medium in which isolated bundle sheath cells prepared from illuminated plants were suspended, also led to a decrease in the starch content after 20 h at 30°C (Table III, experiment B). The analysis of the suspending medium at the end of the incubation time showed that the concentration of glucose phosphates was reduced to one-tenth and that reducing sugars (mainly maltose, but also glucose and oligosaccharides) had appeared. However, the amount of these sugars found was less than one would expect from the sum of glucose-1-phosphate consumed and the starch that disappeared. Here we must point out that the incubation time in the case of the investigation of starch formation from glucose-1-phosphate was 1–2 h whilst in these experiments on starch breakdown we incubated the isolated bundle sheath cells for about 20 h. Therefore, it is conceivable that the carbohydrate lacking in our stoichiometry was being consumed by the cells in order to maintain at least some of their metabolic functions. Porter (1962) discussed such a concurrent production of sugars and CO_2 from starch in the dark.

Our experiments on the formation of starch in isolated bundle sheath cells show that glucose-1-phosphate is readily taken up and that the transfer of glucose to starch is accompanied by the release of inorganic phosphate. It is, therefore, not possible to ascertain whether the effect of glucose-1-phosphate on the stimulation of starch degradation is due to this phosphate *per se* or to an increase in the concentration of inorganic phosphate at the site of starch metabolism as a consequence of the glucose transfer step. Our examination of the starch formation process showed that maltooligosaccharides were produced. This fact indicates the participation of amylases, and all our results suggest that the activities of phosphorylase and amylases are coupled in some way. One would imagine that the stimulation of starch breakdown by fructose-1,6-diphosphate could also be a consequence of a release of inorganic phosphate in the plastids. There are good indications that fructose-1,6-diphosphate is readily transported through the plastid envelope (Heber, 1967; Bassham *et al.*, 1968). As we have mentioned (Table III), less starch is degraded in the detached leaf than in one attached to the plant. We have also observed that detached leaves contain, after 80 h in the light, 2·5 times more sucrose than the leaves on the plant, while the starch content of the detached leaves only increases by about 25% (Vieweg and Fekete, 1972). These two observations suggest that starch content is under control and that the regulating mechanism must be connected with sucrose metabolism.

IV. FINALE: SWEET DREAMS

We have reported (Fekete and Vieweg, 1973a) that bundle sheath preparations from maize leaves, also containing sieve and tracheid elements, are able to synthesize some sucrose if they are supplied with UDPG (or UTP and glucose-1-phosphate) and fructose-6-phosphate or fructose. The amount of sucrose formed is about 10 times less than is produced in the light by leaves attached to plants or, *in vitro*, where leaf strips are incubated with UDPG and fructose-6-phosphate as substrates. These results suggest that sucrose is synthesized in the mesophyll cells. The activities of the sucrose synthesizing enzymes in the bundle sheath cells are also too low to allow all the starch mobilized in the leaves in the dark to be converted to sucrose. From the time dependence of sucrose and starch formation observed in intact maize leaves in the light (Vieweg and Fekete, 1972) we could have concluded that the sucrose formed in the mesophyll was transported to the bundle sheath cells and there metabolized to starch, just as happens in many other plant tissues (cf. Porter, 1962). However, the isolated bundle sheath cells are not capable of converting sucrose to starch. In order to produce starch they depend on the transport of glucose-1-phosphate from the neighbouring mesophyll. Thus the glucose-1-phosphate furnished by photosynthesis is the common metabolite that links sucrose and starch metabolism. Here again, it should be noted that Porter (1962) suggested that a phosphate ester cycle was involved in the concurrent formation of sucrose and starch from CO_2 in photosynthesizing leaves. In maize leaves where there is a distribution of functions, sucrose synthesis occurs in the mesophyll while starch accumulates in the bundle sheath cells. This situation drew our attention to the particular role of glucose-1-phosphate and allowed further clarification of the mechanism of regulation of sucrose and starch metabolism in maize leaves.

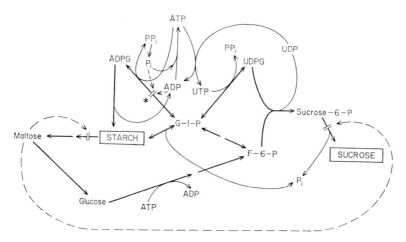

FIG. 5. The main pathways concerned with starch metabolism in maize leaves. Broken lines indicate inhibition of reactions, shown thus ⊣▷, by metabolites. *ADP-glucose pyrophosphorylase is activated by 3-phosphoglycerate.

In Fig. 5 it can be seen glucose-1-phosphate is a substrate shared by phosphorylase, UDPG-pyrophosphorylase and ADPG-pyrophosphorylase and that all of the reactions catalysed by these enzymes can proceed simultaneously if the conditions are suitable. In the light, when photophosphorylation ensures a low concentration of inorganic phosphate (Kandler, 1950), a direct transfer of glucose from glucose-1-phosphate to starch takes place. The activity of phosphorylase in the bundle sheath cells of maize is high enough to account for the starch produced *in vivo* (Fekete and Vieweg, 1971). The formation of ADPG occurs only when the concentration of inorganic phosphate is very low because of the high sensitivity of ADPG-pyrophosphorylase to inhibition by inorganic phosphate (Ghosh and Preiss, 1966). In addition, because of inhibition by ADP, the synthesis of ADPG can only proceed under conditions that ensure the almost complete transformation of inorganic phosphate and ADP to ATP. In turn, ATP is a substrate which is utilized for a number of reactions. One of them leads to the formation of UTP which is required for the synthesis of UDPG. The activity of UDPG-pyrophosphorylase, which catalyses this reaction, is about 100–1000 times higher than that of ADPG-pyrophosphorylase in many plant tissues (Turner, 1969; Fekete, 1969a,b; Huber et al., 1969; Ozbun et al., 1973). Therefore, the synthesis of UDPG is always assured while ADPG is only formed when ATP pools are replenished.

The relative rates of utilization of glucose-1-phosphate by phosphorylase and UDPG-pyrophosphorylase, which are both very active, would depend on the concentration of glucose-1-phosphate at the site where the enzyme is located. Thus, if the K_m value for glucose-1-phosphate in the case of maize leaf UDPG-pyrophosphorylase is similar to that reported by Hansen et al. (1966) for the liver enzyme ($5 \cdot 0 \times 10^{-5}$ M) and if other conditions are suitable, one would expect the formation of UDPG to proceed at maximum rate even at low glucose-1-phosphate concentrations. Phosphorylases have a K_m for glucose-1-phosphate of about 10^{-3} M (Tsai and Nelson, 1969; Rebenich, 1972). Therefore, these enzymes would synthesize starch at an appreciable rate only when the concentration of glucose-1-phosphate is about 10-fold higher than that needed for maximum UDPG formation. This preference for UDPG synthesis with the concomitant production of sucrose can explain the absence or low content of starch in leaves of fast-growing plants where sucrose is rapidly being translocated away from the tissues. On the other hand, when growth is inhibited, starch accumulates (cf. Porter, 1962) because less glucose-1-phosphate is required for UDPG and sucrose synthesis.

Our results suggest that in maize leaves, even during starch synthesis, the amylases are active and their activities are coupled to those of phosphorylase. The amylolytic breakdown of starch produces maltose and other oligosaccharides and these products serve as new primers for further synthesis in the light. Maltose is a good glucosyl acceptor for the reaction catalysed by one of the starch synthetase isozymes (Akazawa, 1972) and the oligosaccharides thus produced can serve as primers for phosphorylase I: the oligosaccharides furnished by α-amylase can also be lengthened by this enzyme. When the oligosaccharides attain a certain chain length, then further chain elongation can be taken over by phosphorylase II and by the synthetase requiring ADPG if synthesis of this sugar nucleotide is possible. Whenever the maltose concentration rises, the activity of amylases is inhibited and further starch breakdown is avoided.

It is interesting that, in our experiments, maltose was further metabolized in leaf strips in the light (Vieweg and Fekete, 1973) but not in the isolated bundle sheath cells (Fekete and Vieweg, 1973b). Therefore, in these latter cells the regulatory role of maltose is more important than its primer function.

It seemed obvious to us that if maltose played such an important role in maize leaves, then it ought to be possible to find some of the disaccharide in leaf extracts. In spite of the statement by Akazawa (1965) that neither maltose nor dextrins are natural constituents in plant cells, our chromatographic examination of methanolic extracts of illuminated maize leaves showed, besides sucrose, significant amounts of maltose, glucose, fructose and oligosaccharides.

Our results suggest that the amylolytic breakdown of starch is also controlled by the presence of the phosphorylase protein. This enzyme is able to produce stable complexes with starch primers of appropriate chain lengths and presumably the bound starch would then not be susceptible to amylase attack. If this is the case one would expect that a certain amount of residual starch would be always maintained in leaves even after many days in the dark. This was confirmed with maize leaves which were found to contain measurable amounts of starch even after 4 days in the dark. Similar results were reported by Komuro and co-workers (1966). With this residual starch tightly bound to phosphorylase the problem of a primerless synthesis is avoided. Whenever the glucose-1-phosphate concentration rises there could be a rapid transfer of glucose residues to the bound polymer which would be independent of the presence or absence of free starch. When the chains attained a certain length, new primers could then be produced by the action of amylases.

In Fig. 5 we can see that inorganic phosphate and maltose may serve as further regulating links between sucrose and starch metabolism. In a system where the concentration of total phosphate is almost constant, its association with different cell constituents may play a regulating role. Thus, if the metabolism of a phosphorylated compound is inhibited and it accumulates, the turnover of the whole system may cease because the bound phosphate is then not available for other reactions. Such a case occurs in leaves when sucrose phosphatase is inhibited and phosphate remains bound to sucrose. Hawker (1967) reported that this enzyme can be inhibited by sucrose and maltose. Our experiments with leaves detached from plants showed that in the dark less starch was degraded while in the light more starch accumulated than in the controls left on the plants. By detaching the leaves the transport of sucrose is prevented and a sucrose accumulation can be demonstrated. The increase of sucrose concentration then presumably produces an inhibition of sucrose phosphatase. Therefore, some of the newly formed sucrose phosphate is not hydrolysed and the concentration of inorganic phosphate diminishes, thus favouring starch synthesis. Also, owing to the high sucrose concentration, less UDP-glucose should be utilized for further sucrose synthesis. Perhaps the reaction catalysed by sucrose synthetase can also contribute to an increase in the UDP-glucose concentration. As a consequence of this more glucose-1-phosphate should be available for further starch synthesis. In the dark a lack of inorganic phosphate would slow down the phosphorolytic breakdown of starch. Further decrease of inorganic phosphate would produce a decrease in the size of the ATP pools, hence maltose could not be metabolized and would accumulate. At higher maltose concentrations inhibition of amylases and sucrose phosphatase may be envisaged and a complete cessation of starch breakdown may occur.

The summation of all the effects that have been discussed could lead to very accurate control of the different pathways. The distribution of a limited amount of phosphate amongst different metabolites means that the concentration of inorganic phosphate reflects the state of the system at any moment. Therefore it is understandable that the inorganic phosphate not only acts as a substrate and product, but also as an effector which influences several enzymes.

ACKNOWLEDGEMENT

This work has been supported by grants from the Deutsche Forschungsgemeinschaft.

REFERENCES

Akazawa, T. (1965). *In* "Plant Biochemistry" (J. Bonner and J. E. Varner, eds), p. 258. Academic Press, New York and London.

Akazawa, T. (1972). *In* "Biochemistry of the Glycosidic Linkage: An integrated View" (R. Piras and H. G. Pontis, eds), p. 305. Academic Press, New York and London.

Badenhuizen, N. P. (1965). *Stärke* **17**, 69.

Badenhuizen, N. P. (1971). *In* "Handbuch der Stärke" (M. Ulmann, ed.), Vol. VI(2), p. 41. Paul Parey, Berlin and Hamburg.

Badenhuizen, N. P. and Chandorkar, K. R. (1965). *Cereal Chem.* **42**, 44.

Bassham, J. A., Kirk, M. and Jensen, R. G. (1968). *Biochim. biophys. Acta* **153**, 211.

Baun, L. C., Palmiano, E. P., Perez, C. M. and Juliano, B. O. (1970). *Pl. Physiol.* **46**, 429.

Chan, T. T. and Bird, I. F. (1960). *J. exp. Bot.* **11**, 335.

Chandorkar, K. R. and Badenhuizen, N. P. (1967). *Cereal Chem.* **44**, 27.

Doi, K. and Doi, A. (1969). *J. Japan Soc. Starch Sci.* **17**, 89.

Ewart, M. H., Siminovitch, D. and Briggs, D. R. (1954). *Pl. Physiol.* **29**, 407.

Fekete, M. A. R. de (1966). *Archs Biochem. Biophys.* **116**, 368.

Fekete, M. A. R. de (1968). *Planta* **79**, 208.

Fekete, M. A. R. de (1969a). *Planta* **87**, 311.

Fekete, M. A. R. de (1969b). *Planta* **87**, 324.

Fekete, M. A. R. de and Cardini, C. E. (1964). *Archs Biochem. Biophys.* **104**, 173.

Fekete, M. A. R. de, Leloir, L. F. and Cardini, C. E. (1960). *Nature, Lond.* **187**, 918.

Fekete, M. A. R. de and Vieweg, G. H. (1971). *Ber. deut. bot. Ges.* **84**, 475.

Fekete, M. A. R. de and Vieweg, G. H. (1973a). *Ber. deut. bot. Ges.* **86**, 227.

Fekete, M. A. R. de and Vieweg, G. H. (1973b). *Ann. N.Y. Acad. Sci.* **210**, 170.

French, D. and Youngquist, R. W. (1963). *Stärke* **12**, 425.

Gafin, J. E. and Badenhuizen, N. P. (1959). *S. Afr. J. Sci.* **55**, 73.

Gates, J. W. and Simpson, G. M. (1968). *Can. J. Bot.* **46**, 1459.

Ghosh, H. P. and Preiss, J. (1966). *J. biol. Chem.* **241**, 4491.

Haapala, H. (1969). *Physiol. Pl.* **22**, 140.

Hansen, R. G., Albrecht, G. J., Bass, S. T. and Seifert, L. L. (1966). *In* "Methods in Enzymology" (S. P. Colowick and N. O. Kaplan, eds), Vol. VIII, p. 248. Academic Press, New York and London.

Hatch, M. D., Osmond, C. B. and Slatyer, R. O. (1971). "Photosynthesis and Photorespiration." Wiley-Interscience, New York, London, Sydney, Toronto.

Hawker, J. S. (1967). *Biochem. J.* **102**, 401.

Heber, U. W. (1967). *In* "Biochemistry of Chloroplasts" (T. W. Goodwin, ed.), Vol. II, p. 71. Academic Press, London and New York.

Huber, W., Fekete, M. A. R. de and Ziegler, H. (1969). *Planta* **87**, 360.

Juliano, B. O. and Varner, J. E. (1969). *Pl. Physiol.* **44**, 886.

Kandler, O. (1950). *Z. Naturf.* **5b**, 423.

Komuro, T., Yano, K. and Hattori, S. (1966). *Bot. Mag. Tokyo* **79**, 376.

Murata, T., Sugiyama, T. and Akazawa, T. (1964). *Archs Biochem. Biophys.* **107**, 92.

Ozbun, J. L., Hawker, J. S., Greenberg, E., Lammel, C. and Preiss, J. (1973). *Pl. Physiol.* **51**, 1.

Porter, H. K. (1962). *A. Rev. Pl. Physiol.* **13**, 303.

Potter, P. K. (1964). "Some aspects of the enzymology of potato tubers. Intracellular distribution of some enzymes of starch and sucrose metabolism." Dissertation, University of Western Australia, Perth, Australia.

Rebenich, P. (1972). "Trennung und Eigenschaften der Phosphorylasen bei Vicia faba." Zulassungsarbeit, Technische Hochschule Darmstadt.

Schwimmer, S. (1950). *J. biol. Chem.* **186**, 181.

Shannon, J. C. (1968). *Pl. Physiol.* **43**, 1215.

Smith, E. E., Taylor, P. M. and Whelan, W. J. (1968). *In* "Carbohydrate metabolism and its Disorders" (F. Dickens, P. J. Randle and W. J. Whelan, eds), Vol. 1, p. 89. Academic Press, London and New York.

Stocking, C. R. (1952). *Am. J. Bot.* **39**, 283.

Tsai, C. Y. and Nelson, O. E. (1969). *Pl. Physiol.* **44**, 159.

Turner, J. F. (1969). *Aust. J. biol. Sci.* **22**, 1321.

Vieweg, G. H. and Fekete, M. A. R. de (1972). *Planta* **104**, 257.

Vieweg, G. H. and Fekete, M. A. R. de (1973). *Ber. deut. bot. Ges.* **86**, 233.

Viswanathan, P. N. and Krishnan, P. S. (1966). *Indian J. Biochem.* **3**, 228.

Ziegler, H. (1956). *Planta* **47**, 447.

CHAPTER 10

The Primary Cell Wall and Control of Elongation Growth*

PETER ALBERSHEIM

*Department of Chemistry, University of Colorado,
Boulder, Colorado, U.S.A.*

I. INTRODUCTION

The reactions which regulate elongation growth of plant cells are unknown but are of great interest. Elongation growth requires the presence of hormones and most researchers who have been concerned with cell elongation have fixed their attention on the parameters affecting the hormone activation of this process. Our approach to this problem has been based on structural analysis of the cell wall. Since the wall has been demonstrated to limit the rate of growth (Cleland, 1971) and since the wall must be chemically altered in order to permit growth, we felt that a true understanding of these growth processes would require the elucidation of the structure of the wall. I shall summarize our current level of understanding of cell wall structure and then speculate on the reactions which take place within the wall during growth and how, biochemically, this growth is controlled by hormones.

Cell walls are conveniently considered to be of two types, thin primary walls and thicker secondary walls. Primary walls are laid down by young, undifferentiated cells that are still growing. Primary walls are transformed into secondary walls after the cell has stopped growing. The primary cell walls of a variety of higher plants appear to have many features in common and may, in fact, have very similar structures. This is not true of secondary walls where the composition and ultrastructure vary considerably from one

* Supported in part by Atomic Energy Commission grant AT(11-1)-1426.

cell type to another. The results I shall report to you come predominantly from studies of walls isolated from suspension-cultured sycamore (*Acer pseudoplatanus*) cells. More recent results indicate that walls isolated from suspension-cultured bean (*Phaseolus vulgaris*) cells are very similar to the walls of the distantly related sycamore cells.

There were two commanding reasons for selecting the cell walls isolated from suspension-cultured cells for structural analysis. The most important is that many plants when grown in suspension-culture exist as a homogeneous tissue possessing primary but no secondary walls. The second important attribute of such cultured cells is that they secrete into their culture medium polysaccharides which are similar in composition to the non-cellulosic polysaccharides of the cell wall (Becker *et al.*, 1964). It has been suggested (Becker *et al.*, 1964) and now confirmed (Bauer *et al.*, 1973) that at least some of these extracellular polysaccharides are structurally related to cell wall polysaccharides. Therefore, these extracellular polymers offer a convenient source of material for developing techniques to study the wall polymers.

The sugar composition of the total wall, of isolated polysaccharides, or of polysaccharide fragments, has been obtained by the formation of alditol acetate derivatives. These volatile compounds may be separated and quantitatively analyzed by gas chromatography and computer-assisted data reduction of the gas chromatographic output. This method allows relatively quick, facile and accurate analysis of sub-milligram quantities of all of the neutral and acidic sugars present in plant cell walls (Jones and Albersheim, 1972).

The structural analysis of cell walls has been facilitated by recent improvements in the technique of methylation analysis. The complete methylation of most polysaccharides can now be achieved in one or at most two methylation reactions by the method of Hakomori (1964). Probably the most important advance in the area of methylation analysis has been the development of an easy procedure for the qualitative and quantitative analysis of the methylated sugars obtained upon acid hydrolysis of methylated polysaccharides (Björndal *et al.*, 1970). In this procedure, the partially methylated sugars are converted to the *O*-methyl alditols; these are then acetylated. The resulting partially methylated alditol acetates are separated, identified and quantitatively estimated by the combined application of gas chromatography and mass spectrometry. The mass spectrometric identifications are based on the fact that a unique fragmentation pattern is obtained for each methyl substitution pattern in the partially methylated alditol acetates. These mass spectrometric data, in combination with relative retention time data obtained from several gas chromatographic columns, will, in most cases, give sufficient evidence for an unambiguous identification of the methylated sugar. The major advantage of this procedure is that crystalline derivatives are not necessary for the identification of the methylated sugars and a methylation analysis can be performed accurately on milligram quantities of material.

Although the general components of plant cell walls have been identified and studied in considerable detail, only recently has a picture of the molecular structure of a plant cell wall been presented. No such a picture could be drawn until the general components of the wall were resolved into distinct polymers of definite composition, linkage, size and sequence, and until the interconnections between these polymers were identified (Bauer *et al.*, 1973; Keegstra *et al.*, 1973; Talmadge *et al.*, 1973). The major difficulty in any analysis of cell wall structure has been finding suitable methods for the isolation of *defined* fragments from the wall. Classical methods of extraction, using acid or base, resulted in the simultaneous, but partial, cleavage of several types of bonds present in the plant cell wall. This lack of specificity made it most difficult to determine how the fragments extracted with acid or base were linked to each other or to the residual wall structure. A much more satisfactory fragmentation of the cell wall has been achieved through the use of purified hydrolytic enzymes. The specificity of a given hydrolytic enzyme can be determined with model substrates, and with this knowledge, the linkages between enzymically released wall fragments and the residual wall may be reconstructed.

Those plant pathogens that have the capacity to degrade the cell walls of their hosts represent an excellent source of the highly specific polysaccharide-degrading enzymes required for the structural analysis of cell walls (Albersheim *et al.*, 1969; Bateman *et al.*, 1969). The fungal pathogen *Colletotrichum lindemuthianum* has been shown to secrete different enzymes in a temporal sequence with regard to culture age (English and Albersheim, 1969). The first enzyme secreted by this fungus is an endopolygalacturonase. This enzyme has been highly purified (English *et al.*, 1972) and is able to remove much of the galacturonic acid from the isolated cell walls of a number of plants. The purified endopolygalacturonase has been very important in our present studies as it is the only enzyme which, by itself, has been found to initiate effectively the degradation of isolated cell walls.

II. THE CELL WALL HEMICELLULOSE

Let us examine briefly the structure of a xyloglucan, the only hemicellulose present in suspension-cultured sycamore walls. The structural analysis of this important wall polymer is also a fine example of the efficacy of purified polysaccharide-degrading enzymes when combined with gas chromatographic-mass spectrometric analysis. The gas chromatogram in the top of Fig. 1 shows the partially methylated alditol acetates derived from this polymer (Bauer *et al.*, 1973). The inositol in this chromatogram has been added as an internal standard. Some of the peaks contain more than a single component; these have been separated by gas chromatography on a different column. Each of the components in the gas chromatogram, including the arabinose,

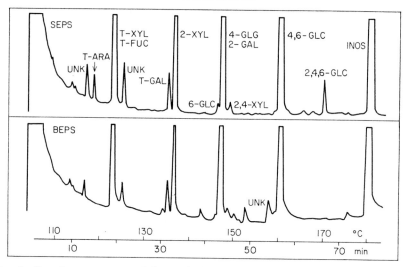

Fig. 1. Gas chromatograms of the methylated alditol acetates obtained from purified xyloglucan of sycamore (top) (Bauer *et al.*, 1973) and of bean (bottom) (Wilder and Albersheim, 1973). The initial peak in the chromatograms is due to the acetic anhydride used as the solvent. The glycosidic linkages to each sugar derivative are indicated by numerical prefixes: thus, 4,6-GLC indicates that sugars are glycosidically linked in the polysaccharide to carbons 4 and 6 of glucosyl residues. Terminal residues are indicated by T- (e.g., T-XYL). The abbreviations for the sugars are: arabinose, ARA; fucose, FUC; galactose, GAL; glucose, GLC and xylose, XYL; and *myo*-inositol, INOS, which was used as a standard. Unidentified components are indicated (UNK).

fucose and galactose derivatives, are part of the polymer which we call xyloglucan.

A truly beautiful example of conservation of structure during evolution was obtained by Barry Wilder (Wilder and Albersheim, 1973) in my laboratory when he compared the gas chromatograms of the partially methylated alditol acetates obtained from the purified xyloglucan of suspension-cultured bean cells (Fig. 1, bottom) with the chromatography of the same derivatives of the xyloglucan of the very distantly related sycamore cells (Fig. 1, top). This near identity, in addition to many other similarities in the wall structure of these two plants, suggests that the primary cell walls of all flowering plants are likely to be very similar.

The detailed molecular structure of sycamore xyloglucan, as well as of bean xyloglucan, has been characterized by methylation analysis of the oligosaccharides obtained by endoglucanase treatment of the polymer. This enzyme hydrolyzes the glycosidic bonds of glucosyl residues that have another sugar attached to carbon 4, but the enzyme does not hydrolyze the glycosidic bonds of glucosyl residues that have other sugars attached to both carbons 4 and 6 (see Fig. 5). The fragments produced by endoglucanase treatment were separated by chromatography on Bio-Gel P-2 (Fig. 2). Each fragment has been structurally analysed (the analysis of peaks 3 and 4 are given in Figs 3

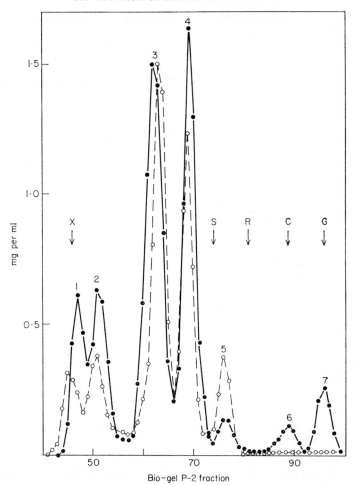

FIG. 2. Bio-Gel P-2 fractionation of endoglucanase-treated sycamore xyloglucan. Untreated xyloglucan voids the column. The elution volumes of untreated xyloglucan (X), stachyose (S), raffinose (R), cellobiose (C), and glucose (G) are indicated by arrows.

and 4) and the structure of the polymer was, in large part, deduced from this information (Bauer *et al.*, 1973). The structure of the xyloglucan is based on a repeating heptasaccharide unit which consists of 4 residues of β-(1 → 4)-linked glucose and 3 residues of terminal xylose. A single xylose residue is glycosidically linked to carbon 6 of three of the glucosyl residues. The xyloglucan appears to consist predominantly of repeating units of the seven sugar fragment in Fig. 3 and the nine sugar fragment in Fig. 4. This latter fragment contains, in addition to the basic heptasaccharide repeating unit in Fig. 3, a fucosylgalactose disaccharide attached to one of the xylose residues. It has not been determined to which of the three xylose residues

	COMPOSITION		LINKAGE	COMPOSITION	
	EXPT	CALC		EXPT	CALC
XYL	44.1	42.9	T−XYL	41.3	42.9
GLC	55.9	57.1	4−GLC	12.7	14.3
			6−GLC	13.4	14.3
			4,6−GLC	27.3	28.6

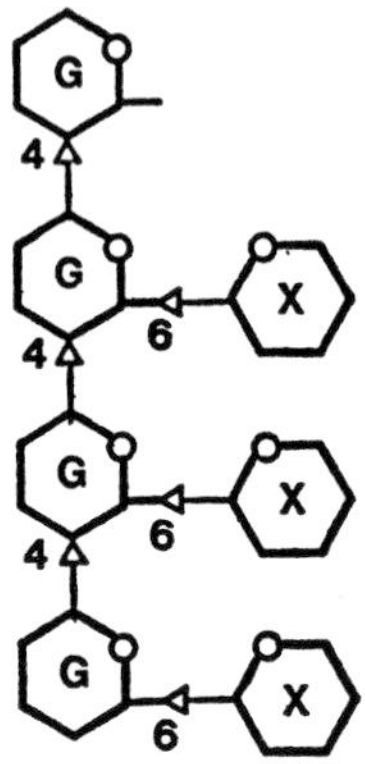

Fig. 3. The structure of the endoglucanase-derived xyloglucan oligosaccharide which elutes as peak 4 from the Bio-Gel P-2 column depicted in Fig. 2 (Bauer *et al.*, 1973). The alditol acetate-determined sugar composition and the partially methylated alditol acetate-determined linkage composition are compared to the values anticipated for the proposed structure. The abbreviations for the sugars are: glucose, GLC or G and xylose, XYL or X.

(in the heptasaccharide repeating unit) the disaccharide is attached. A portion of a xyloglucan molecule is depicted in Fig. 5.

The function of the xyloglucan appears to be based on the ability of its β-(1 → 4)-linked glucan backbone to hydrogen bond to the β-(1 → 4)-glucan chains of cellulose fibers. Models of the xyloglucan indicate that the fucosyl-galactose side chains prevent further β-(1 → 4)-glucan chains from hydrogen bonding to those xyloglucan chains which are already hydrogen-bonded to cellulose fibers. This results in the cellulose fibers being coated with a single layer of xyloglucan chains. It appears that there is sufficient xyloglucan in suspension-cultured sycamore cell walls to coat completely all of the cellulose fibers. The reducing end of many of the xyloglucan chains is covalently attached to the pectic polymers. Thus, the xyloglucan chains serve as a bridge between the cellulose fibers and the rest of the cell wall.

III. The Cell Wall Pectic Polymers

The walls of suspension-cultured sycamore cells contain a single acidic pectic polymer (rhamnogalacturonan) and three neutral pectic polymers

	COMPOSITION		LINKAGE	COMPOSITION	
	EXPT	CALC		EXPT	CALC
FUC	10.3	11.1	T—FUC	11.1	11.1
XYL	33.0	33.3	T—XYL	20.4	22.2
GAL	12.1	11.1	2—XYL	11.3	11.1
GLC	45.2	44.4	T—GAL	2.3	0
			2—GAL	9.4	11.1
			4—GLC	12.8	11.1
			6—GLC	10.8	11.1
			4,6—GLC	21.1	22.2

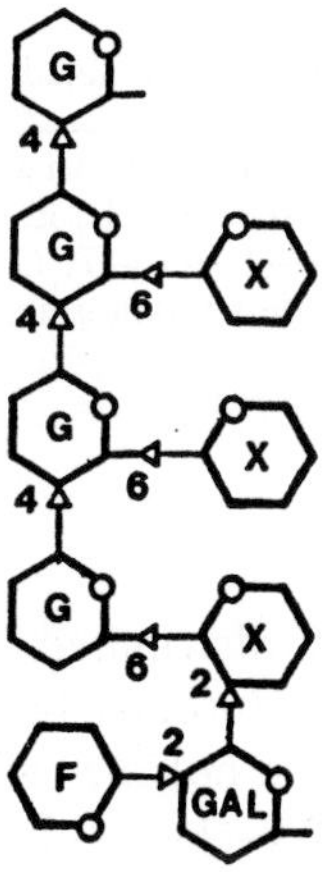

FIG. 4. The structure of the endopolyglucanase-derived xyloglucan oligosaccharide which elutes as peak 3 from the Bio-Gel P-2 column as depicted in Fig. 2 (Bauer *et al.*, 1973). The alditol acetate-determined sugar composition and the partially methylated alditol acetate-determined linkage composition are compared to the values anticipated for the proposed structure. The particular xylose residue to which the fucosylgalactose is attached is unknown. The abbreviations for the sugars are: fucose, FUC or F; galactose, GAL; glucose, GLC or G and xylose, XYL or X.

(β(?)-(1 → 4)-galactan, branched arabinan, and (3 → 6)-linked arabinogalactan). A tentative structure for the rhamnogalacturonan of sycamore walls is presented in Fig. 6. This structure is based on the analyses of fractions obtained by endopolygalacturonase treatment of cell walls and on the analysis of the galacturonosyl-containing oligomers obtained by partial acid hydrolysis of isolated cell walls. The rhamnogalacturonan is not a straight chain molecule. The zigzagged shape results from the presence of 2-linked rhamnosyl residues in an otherwise α-(1 → 4)-linked galacturonan chain. When another sugar is attached to carbon 4 of a rhamnosyl residue, the rhamnose forms a Y-shaped branch point. These observations came as a result of building CPK space filling models of these structures. Similar observations have been made by Rees and Wight (1971) working with models of pectin and by

XYLOGLUCAN

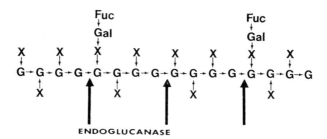

FIG. 5. A proposed structure for a portion of a xyloglucan molecule from sycamore cell walls. The structure is based on the evidence presented by Bauer *et al.* (1973). The number of glucosyl residues in a complete molecule is thought to be between 25 and 30. The abbreviations for the sugars are: fucose, FUC; galactose, GAL; glucose, G and xylose, XYL. The glycosyl linkages are G-β-(1 → 4)-G; X-α-(1 → 6)-G; Gal-β-(1 → 2)-X; and Fuc-α-(1 → 2)-Gal. The glycosidic linkages susceptible to endoglucanase hydrolysis are indicated by arrows.

Simmons (1971) in studying the *O*-antigen polysaccharides of *Shigella flexneri*.

The rhamnosyl residues are not randomly distributed in the chain, but probably occur as rhamnosyl-(1 → 4)-galacturonosyl-(1 → 2)-rhamnosyl units. This sequence appears to alternate in the polymer with a homogalacturonan sequence containing approximately 8 α-(1 → 4)-linked galacturonic acid residues. These two sequences give rise to the two major fractions obtained from endopolygalacturonase treatment of sycamore cell walls. One fraction consists of a mixture of mono-, di-, and tri-galacturonic acid and arises from the endopolygalacturonase hydrolysis of the homogalacturonan sequences of the rhamnogalacturonan polymer. The second fraction contains neutral sugar-rich polymers and arises from the rhamnose-rich region of the rhamnogalacturonan polymer.

DEAE-Sephadex ion exchange chromatography demonstrates that the neutral sugars of the endopolygalacturonase products are covalently attached to galacturonosyl residues. Most of these neutral sugar residues, which represent 76% of this fraction, are part of either a branched arabinan or a linear (1 → 4)-galactan. Methylation analysis has shown that approximately 50% of the rhamnosyl residues in the wall are branched, having a substituent at carbon 4 as well as a galacturonosyl residue attached to carbon 2. As this is the major branch point of the rhamnogalacturonan chain, the (2 → 4)-linked rhamnosyl residue is likely to represent the point of attachment of either the (1 → 4)-galactan or the branched arabinan or of both of these polymers.

Most of the arabinosyl and galactosyl residues that are released by endopolygalacturonase treatment are part of acidic polymers. The structure of

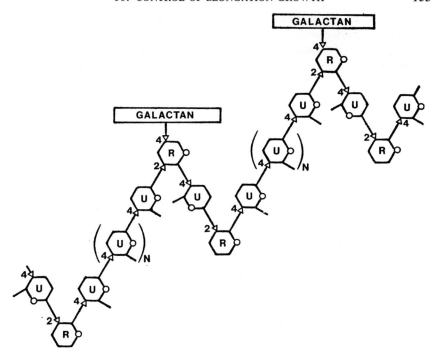

FIG. 6. A proposed structure for the rhamnogalacturonan of sycamore cell walls. The structure is based on evidence presented by Talmadge *et al.* (1973). The number of residues in the homogalacturonan regions is unknown, but "N" is estimated to be about 6. The abbreviations for the sugars are: rhamnose, R and galacturonic acid, U.

these polymers was examined by partial acid hydrolysis. Following such hydrolysis, the sample was fractionated on DEAE-Sephadex into neutral and acidic components which were then separately chromatographed on a Bio-Gel P-2 column. The results from this analysis indicate that about 85% of the arabinosyl residues were released by the partial hydrolysis from the acidic rhamnogalacturonan fragment while 75% of the galactosyl residues remain attached to the acidic fragment. This preferential cleavage of arabinose is accompanied by only minor changes in the linkages of the sugar residues remaining in the acidic fraction. These results in conjunction with methylation analysis, which showed the presence of only small amounts of branched galactosyl residues and large amounts of terminal and branched arabinosyl residues, suggest that the galactosyl residues are present as a linear chain which is attached at its reducing end to the rhamnogalacturonan main chain, and that the arabinosyl residues are in the form of a branched chain. Recent results (McNeil and Albersheim, unpublished) suggest that the arabinan is attached directly to the rhamnogalacturonan rather than to the galactan.

    The endopolygalacturonase products contain, in addition to the arabinan, the $(1 \rightarrow 4)$-linked galactan, and the rhamnogalacturonan fragment, small amounts of xyloglucan. The xyloglucan of the endopolygalacturonase-liberated polymers fractionates as an acidic polymer on DEAE-Sephadex, indicating that there is a covalent attachment between the neutral xyloglucan chains and the acid pectic polysaccharides. After weak acid hydrolysis of this portion of the cell wall, over 70% of the xyloglucan still fractionates as an acidic polymer on DEAE-Sephadex. This indicates that the xyloglucan may be attached to the rhamnogalacturonan via the galactan; and this result is further evidence that acid-labile arabinosyl residues are not interspersed in the galactan chain. Thus, the galactan may serve as a cross-link between the xyloglucan and rhamnogalacturonan components of the wall.

The structures proposed for the sycamore pectic polymers are generally consistent with previous structural studies on pectic polysaccharides. Rhamnogalacturonans appear to be a common feature of all pectic poly-saccharides. Pectic galactans containing a β-$(1 \rightarrow 4)$-linked galactan backbone have been isolated from soybean seed (Aspinall et al., 1967). The highly branched arabinan region of sycamore pectic polysaccharides contains similar linkages to those present in the pectic arabinans isolated from soybean, lemon peel, and mustard seed (Aspinall and Cottrell, 1971).

Another wall polysaccharide has been identified which contains both arabinosyl and galactosyl residues. This arabinogalactan is only questionably listed as a pectic polymer, and our evidence suggests that it may function as a covalent link between the rhamnogalacturonan chains and the hydroxypro-line-rich wall protein. Most of our work with this polymer has been carried out with the arabinogalactan isolated from the extracellular polysaccharides of suspension-cultured sycamore cells (Keegstra et al., 1973). The arabino-galactan, at pH 2, absorbs to a SE-Sephadex column and, when such a column is eluted with a linear salt gradient, the arabinogalactan elutes simultaneously with a hydroxyproline-containing protein. This suggests that the arabino-galactan and the protein are parts of the same molecule.

Methylation analysis indicates that the arabinogalactan possesses a highly branched structure containing predominantly $(3 \rightarrow 6)$-linked galactosyl residues as branch points with single arabinosyl residues as the most prevalent side chain. These results are similar to those reported by Aspinall et al. (1969) for this arabinogalactan and are also similar to arabinogalactans isolated from coniferous woods (Timell, 1965) and from plant gums (Aspinall, 1969). The sycamore extracellular arabinogalactan differs from those in wood in that it contains a rhamnosyl residue and a higher percentage of arabinose. An arabinogalactan isolated from maple (Acer saccharum) sap is similar to the sycamore (Acer pseudoplatanus) arabinogalactan in all respects (Adams and Bishop, 1960).

One possible structure for the arabinogalactan from sycamore is shown in Fig. 7. This structure has two features which suggest that it is an interesting cell wall component. The first is the attachment of this polysaccharide to a

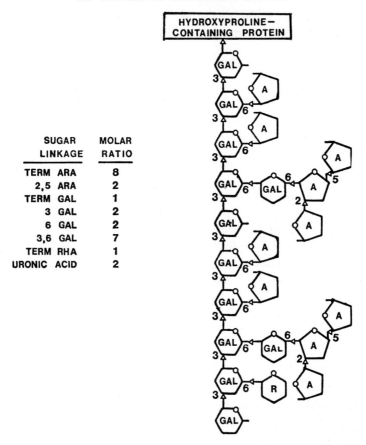

SUGAR LINKAGE	MOLAR RATIO
TERM ARA	8
2,5 ARA	2
TERM GAL	1
3 GAL	2
6 GAL	2
3,6 GAL	7
TERM RHA	1
URONIC ACID	2

FIG. 7. The molar ratio of the glycosyl derivatives present in sycamore extracellular arabinogalactan and a proposed structure for this polysaccharide. The molar ratios were determined by Keegstra *et al.* (1973). Since uronic acids are not recovered from the methylation analysis under the conditions used, it was not possible to tell how the uronic acids are linked in this structure. The structure shown is not unique to the data, but it is consistent with and accounts for the data available. This polymer may be connected to the serine residues of the hydroxyproline-rich structural wall protein. The abbreviations for the sugars are: arabinose, A; galactose, GAL and rhamnose, R.

hydroxyproline-containing protein, a known component of plant cell walls (Lamport and Miller, 1971; Lamport, 1970). Hence, this arabinogalactan may be a wall component, and it may be a connecting point between wall polysaccharides and wall protein. Recent evidence from Lamport's laboratory (Lamport *et al.*, 1973) as well as from our laboratory suggests that the arabinogalactan may be attached to the hydroxyproline-rich protein by linkage to the hydroxyl groups of serine residues. The second striking feature of the arabinogalactan is the presence of a terminal rhamnosyl residue. Since all of the rhamnose in the cell wall is thought to be covalently linked in the rhamnogalacturonan (Talmadge *et al.*, 1973), the rhamnosyl residue in the arabinogalactan might act as a primer to which a rhamnogalacturonan chain is attached.

IV. A TENTATIVE MOLECULAR STRUCTURE OF SYCAMORE CELL WALLS

Direct methylation analysis of the isolated but unfractionated sycamore cell wall has been important in our studies for it yielded a quantitative summary of all but the minor sugar linkages present in the complete wall (Talmadge *et al.*, 1973). Subsequent studies on wall fractions obtained by the action of purified hydrolytic enzymes were made quantitative by comparing the amounts of the sugar linkages found in each fraction to the amounts present in the whole wall. These results demonstrate that the sycamore cell wall is composed of a limited number of major structural components: a branched arabinan, cellulose, $(1 \rightarrow 4)$-linked galactan, the hydroxyproline-rich glycoprotein, rhamnogalacturonan, xyloglucan, and a small amount of 3,6-linked arabinogalactan (Table I). Essentially the entire cell wall is accounted for by

TABLE I

Polymer composition of suspension-cultured sycamore cell walls

Wall component	Cell walls
	%
Arabinan	10
3,6-linked arabinogalactan	2
4-linked galactan	8
Cellulose	23
Protein	10
Rhamnogalacturonan	16
Tetra-arabinosides (attached to hydroxyproline)	9
Xyloglucan	21
Total	99

the sum of the amounts of these seven polymeric components. The impression we had when we started this study was that the plant cell wall was made up of a relatively large number of complex polymers. The fact that the sycamore cell wall is composed of a limited number of well-defined structural components is a finding of utmost importance to plant cell wall research. Future studies of other plant cell walls and of the structural changes which take place during growth and development will be encouraged and facilitated by this knowledge.

A model (Fig. 8) of the sycamore cell wall can be proposed from our results (Keegstra *et al.*, 1973). Although other structures are possible, the one presented is consistent with all of the data obtained. The model utilizes

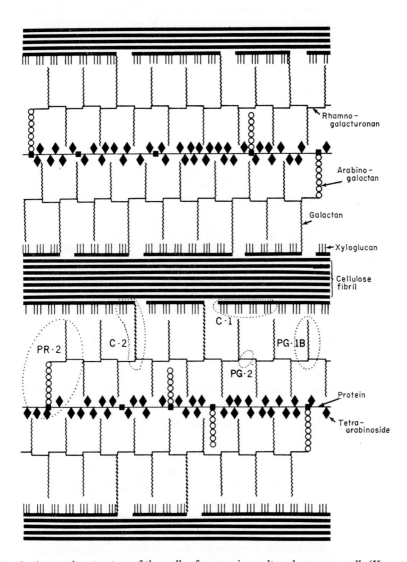

FIG. 8. A tentative structure of the walls of suspension-cultured sycamore cells (Keegstra *et al.*, 1973). The structure presented is discussed in the text. This model is not intended to be quantitative, but an effort has been made to present the wall components in approximately correct proportions. The distance between cellulose fibers is expanded to allow room to present the interconnecting structure. There are probably between 10 and 100 cellulose fibers across a single primary cell wall.

the fact that xyloglucan has been shown to bind tightly to purified cellulose (Aspinall *et al.*, 1969) as well as to the cellulose of the cell wall (Bauer *et al.*, 1973). Several lines of experimental evidence (Keegstra *et al.*, 1973) indicate that the reducing ends of the xyloglucans are attached to the galactan side chains of the rhamnogalacturonan (Fig. 8). A single pectic polysaccharide is likely to be attached through xyloglucan chains to more than one cellulose fiber; and a single cellulose fiber attached through xyloglucan chains to more than one pectic polysaccharide. Such an arrangement would result in cross-linking of the cellulose fibers. The hypothetical structure presented in Fig. 8 suggests, too, that the pectic polysaccharides are attached to seryl residues of the wall protein through a 3,6-linked arabinogalactan chain (Keegstra *et al.*, 1973). This connection would also result in cross-linking of the cellulose fibers. The wall is likely to possess both modes of cross-linking.

The structure presented in Fig. 8 allows an understanding of the wall fragments that each enzyme releases. An example of that portion of the wall solubilized by each enzyme is circled and labeled. These fractions are summarized as follows. The endopolygalacturonase attacks the galacturonosyl linkages of the main pectic chain releasing tri-, di, and mono-galacturonic acids (Fraction PG-2) as well as arabinan and galactan side chains attached to acidic fragments of the main rhamnogalacturonan chain (Fraction PG-1B). After the pectic polysaccharide has been partially degraded by the endopolygalacturonase, endoglucanase more readily degrades xyloglucan, releasing neutral oligosaccharides (Fraction C-1), as well as pectic fragments that are made insoluble by their connection with xyloglucan (Fraction C-2). Pronase, which cannot release significant amounts of carbohydrate from untreated walls, is able to release pectic fragments not released by endopolygalacturonase pretreatment, and larger amounts of carbohydrate are released by pronase after a combination of endopolygalacturonase and endoglucanase treatments (Fraction PR-2).

The primary cell wall of sycamore cells can be considered as a single macromolecule (Fig. 8). The components of the cell wall are, with the exception of the connection between xyloglucan and cellulose, interconnected by covalent bonds; the many hydrogen bonds which interconnect cellulose and xyloglucan make this connection as strong as a covalent bond. It has been suggested (Lamport, 1970) that the plant cell wall contains a protein-glycan network analogous to the peptido-glycan network of bacterial cell walls (Ghuysen, 1968). Our results support this analogy.

V. THE PRIMARY CELL WALLS OF OTHER PLANTS

Many features of the sycamore cell wall are found in the cell walls of other plants. The relative amounts of the wall accounted for by the pectic polymers (arabinan, arabinogalactan, galactan, and rhamnogalacturonan), hemicellulose (xyloglucan), cellulose, and protein, as determined by methylation

analysis and enzymic fractionation, agree closely with the previously published values obtained by chemical fractionation (Lamport, 1965; Roelofsen,
1959). Hydroxyproline-containing proteins with their associated oligo-
arabinosides are widespread in the plant kingdom (Lamport, 1970; Lamport
and Miller, 1971). Kooiman (1960) and others have demonstrated that
xyloglucans are present in the cell walls of the cotyledons or endosperms of
a wide variety of plants. There is also a report which provides some evidence
of a connection between the xyloglucan and pectic polysaccharides of the
cell walls of mustard cotyledons; this report describes a pectic polysaccharide
which has been purified to a state that, "if not homogeneous, consists of a
family of related species" (Rees and Wight, 1969). Methylation analysis was
used to demonstrate that this preparation contained xyloglucan as well as
the pectic polymers. Although Rees and Wight (1969) considered the xyloglucan to be a contaminant, we interpret their data as evidence in support
of a covalent linkage between these wall components. We (Wilder and
Albersheim, 1973) have now established that the xyloglucan of suspension-
cultured bean cells is very similar to the structure of the xyloglucan of suspension-cultured sycamore cells. Also, the manner in which the xyloglucan chains
are bound within the walls of bean cells is similar, if not identical, to that of
the sycamore cell wall. This is established by the fact that the extraction from
the bean cell walls of the xyloglucan molecules requires previous treatment of the walls by an endopolygalacturonase just as extraction of the
xyloglucan polymer from sycamore cell walls requires pretreatment with this
enzyme.

An interesting observation concerning the structure of the plant cell walls
has been reported by Grant *et al.* (1969). They have isolated a soluble mucilage
particle from mustard seedlings and have speculated that this particle may
represent a structural unit of the cell wall. The particle consists of a cellulose
fiber encapsulated by other polysaccharides. The composition of the encap-
sulating polysaccharides suggests that they are xyloglucan and pectic polymers.
Thus, the "cell wall unit" of mustard seedlings may be similar to the structure
of the cell walls of the distantly related sycamore tree.

Beans (*Phaseolus vulgaris*) are phylogenetically unrelated to cells of
sycamore (*Acer pseudoplatanus*) although both are dicotyledonous plants.
A comparison of the structures of the cell walls isolated from these suspension
cultures should go far in deciding whether our model is a general one for all
dicot primary cell walls. Consequently, we have established cell-suspension
cultures of beans. Our first results (Wilder and Albersheim, 1973) indicate
that the structure of the walls of these plants is very similar to that of sycamore.

The evidence available in the literature and that reported here strongly
sustains the hypothesis that the interrelationship between the structural
components of the primary cell walls of all higher plants is comparable.
Many details of the cell wall structure remain to be elucidated, yet the model

available allows one to begin asking questions about the important physiological functions of the cell wall. We are particularly interested in examining
the wall for the molecular mechanisms underlying elongation growth.

VI. A HYPOTHETICAL MECHANISM FOR CELL ELONGATION

The goal of many workers in cell wall research is an elucidation of the
mechanism underlying control of cell wall extension. Cleland (1971) has
lucidly summarized the current thinking about wall extension. It is generally
agreed that addition of auxin to tissues deficient in this hormone quickly
causes the primary cell walls of the tissue to be loosened or weakened, such
that the rate of cell extension is increased. Perhaps not as widely held, but
nevertheless accepted by us, is the view that auxin initiates wall loosening
so quickly that *de novo* protein synthesis and *de novo* polysaccharide synthesis
cannot participate in this initiation. Thus, we have examined our structural
model of the primary cell wall with the idea that initiation of wall extension
probably results from a rearrangement or alteration of the existing wall
structure. Regardless of the mechanism involved, since walls grow throughout
their length, in order for a wall to elongate the cellulose fibers within the wall
must be able to slide along their length relative to each other. Examination
of the wall structure suggests an attractive mechanism by which the cellulose
fibers could move relative to one another. The only non-covalent cross-link
between the structural polymers of the wall is the hydrogen bond-mediated
connection between the xyloglucan chains and the cellulose fibers. Extension
of the wall would result if the xyloglucan chains and the cellulose fibers
moved relative to one another.

Movement of the xyloglucan chains along the cellulose fibers could be
accomplished by hydrogen bond "creep", that is, by the xyloglucan moving
along the cellulose fiber like an inchworm. Such a mechanism would only
require the simultaneous breaking of about four consecutive hydrogen bonds.
This possibility is highly feasible, as is demonstrated by the relatively weak
binding of small xyloglucan fragments to cellulose (Bauer *et al.*, 1973).

The rate at which the xyloglucan chains move along the surface of the
cellulose fibers should be increased by conditions which weaken hydrogen
bonds or conditions which increase the rate at which the hydrogen bonds
between xyloglucan are formed and broken. High hydrogen ion concentrations and elevated temperatures are two treatments which could affect this
bonding and increase the rate of "creep". An enhanced rate of "creep"
resulting from low pH or high temperatures would be apparent when dead
as well as live tissues are placed under tension. The enhanced rate of "creep"
in such tissues would be completely and immediately reversed by raising the
pH or lowering the temperature. In addition, it is easy to conceive of undirec-
tional "creep" in walls that are under tension, for the xyloglucan chains
are covalently linked at their reducing ends to the non-cellulosic wall matrix;

therefore, when neighboring cellulose chains move in opposite directions, the xyloglucans on each would "creep" in only one direction.

The possibility that xyloglucan plays a central role in extension growth has been greatly enhanced by a completely independent finding. John Labavitch and Peter Ray at Stanford University have found that indoleacetic acid stimulates the turnover of the solubilization of xyloglucan from pea stem sections (personal communication). This effect is observed soon after application of auxin (15 min) and confined, among the various wall polymers, to xyloglucan. The finding of xyloglucan in pea stems strengthens the hypothesis that hemicelluloses with similar properties are present in all primary cell walls.

It is possible to speculate on how Labavitch's and Ray's observed rapid turnover of xyloglucan integrates with our own proposed role for this polymer. One explanation that agrees with the observations of both laboratories is that some of the xyloglucan chains are stimulated by auxin to "creep" free or almost free of the cellulose fibers. Since about 65% of the xyloglucan chains of sycamore cells are not covalently attached to the pectic polysaccharides, analogous chains might be found in the buffer extract of pea stems. In any case, the independent acquisition of evidence in two laboratories that xyloglucan may be involved in extension growth gives high priority to further investigation of this polymer.

In our laboratory we have begun to investigate the binding of xyloglucan to cellulose. It is not possible to examine the rate at which intact xyloglucan chains bind and release from cellulose as the binding between these polymers is irreversible under most conditions. Therefore, we have worked with the 7- and 9-sugar fragments released from xyloglucan by the action of endoglucanase and purified by gel filtration chromatography on Bio-Gel P-2 (these are peaks 4 and 3, respectively, from the Bio-Gel P-2 column described by Bauer *et al.* (1973)). Radioactive xyloglucan fragments have been synthesized in order to measure more easily the binding of the fragments to the cellulose fibers. In aqueous solution, these fragments do not bind significantly to cellulose. This is not surprising as a maximum of three hydrogen bonds can form between one fragment molecule and cellulose. However, it is possible to increase the affinity of the xyloglucan fragments for cellulose by reducing the water activity of the solvent. For example, in 75% ethanol and at room temperature, approximately 50% of these fragments bind to cellulose. The bound fragments can be eluted from the cellulose with water. The degree of binding is decreased at higher temperatures and increased at lower temperatures, as would be predicted for a hydrogen bond mode of attachment. The fact that organic solvents such as ethanol and acetone increase the binding of the xyloglucan fragments to cellulose is also expected if these two carbohydrates are held together by hydrogen bonds. The effect of varying the hydrogen ion concentration on the bonding of the xyloglucan fragments to cellulose is in the process of being examined. We recognize that xyloglucan "creep" is only one of many possible mechanisms by which the cell wall can be altered to permit elongation growth.

The ability of hydrogen ions to mimic auxin-induced growth of higher plant cells has been known for some time. Bonner (1934) reported that the growth of oat coleoptile segments was 8 times greater at pH 4·1 than at pH 7·2. This has been confirmed by Nitsch and Nitsch (1956). More detailed knowledge concerning this reaction has recently been obtained by Rayle and Cleland (1970) and by Evans et al. (1971). These workers have found that the low pH response closely resembles the auxin response in a number of ways, including the production of the same maximal growth rate and the necessity for continued presence of the inducing agent for continued rapid growth. The similarity in the growth response to low pH and to auxin suggests the possibility that these responses may share a common mechanism. It has been proposed (Cleland, 1971) that the role of auxin is to activate a hydrogen ion pump in the cell membrane, thereby lowering the pH within the cell wall.

The available evidence supports the idea that an auxin-regulated hydrogen ion pump may be present in plant cell membranes. The presence of a hydrogen ion pump has already been demonstrated in organelle membranes such as those of mitochondria (Mitchell, 1965) and chloroplasts (Jagendorf and Uribe, 1966; Karlish and Avron, 1968), and even in the plasma membranes of animal cells (Nakazawa et al., 1970). Thus, it is not surprising to find evidence for a hydrogen ion pump in plant plasma membranes. Kitasato (1968) has demonstrated a hydrogen ion pump in the plasma membrane of the green alga Nitella, and Jaffe (1970) has presented evidence for the presence of a hydrogen ion pump in the plasma membranes of the root tip cells of mung beans (Phaseolus aureus). The results of Hager et al. (1971) suggest that an auxin-regulated hydrogen ion pump may be present in the cell membranes of oat coleoptiles. Their results are based on the observation that carbonylcyanide-m-chlorophenylhydrazone (CCCP), which is known to make membranes permeable to protons (Hager, 1969), rapidly inhibits auxin-induced elongation of coleoptile segments.

Several workers believe that auxin interacts with the plasma membrane (Brauner and Diemer, 1967; Etherton, 1970; Stowe and Dotts, 1971; Weigl, 1969). Such an interaction might be expected if auxin has the ability to activate a membrane-contained hydrogen ion pump. A test for an auxin-regulated hydrogen ion pump in plasma membranes would consist of measuring the hydrogen ion efflux from cells in the presence as well as in the absence of auxin. An auxin-regulated hydrogen ion pump would be implicated if the hydrogen ion efflux is greater in the presence of physiological concentrations of auxin than in the absence of auxin. Mina Fisher (personal communication) in my laboratory has recently demonstrated very active hydrogen ion pumps in the plasma membranes of suspension-cultured sycamore and bean cells. These pumps are energy-requiring, calcium ion-dependent systems. So far, it has not been possible to demonstrate that the hydrogen ion pumps require the presence of auxin as we have not been able to obtain cells free of endogenous auxin.

One experiment which we intend to carry out in the future is to examine the suspension-cultured cell plasma membranes for the presence of a calcium ion-dependent ATPase. Such an enzyme might be activated by auxin. This would be analogous to vitamin D activation of calcium ion-dependent ATPase in the brush borders of both chick and rat intestine (Martin *et al.*, 1969; Melancon and DeLuca, 1970). If auxin can activate a hydrogen ion pump in the plasma membrane and thereby reduce the pH of the cell wall, one has a logical explanation for the manner in which this hormone controls elongation growth.

REFERENCES

Adams, G. A. and Bishop, C. T. (1960). *Can. J. Chem.* **38**, 2380.
Albersheim, P., Jones, T. M. and English, P. D. (1969). *A. Rev. Phytopath.* **7**, 171.
Aspinall, G. O. (1969). *Adv. Carbohyd. Chem.* **24**, 333.
Aspinall, G. O., Begbie, R., Hamilton, A. and Whyte, J. N. C. (1967). *J. chem. Soc.* (C), 1065.
Aspinall, G. O. and Cottrell, I. W. (1971). *Can. J. Chem.* **49**, 1019.
Aspinall, G. O., Molloy, J. A. and Craig. J. W. T. (1969). *Can. J. Biochem.* **47**, 1063.
Bateman D. F., Van Etten, H. D., English, P. D., Nevins, D. J. and Albersheim, P. (1969). *Pl. Physiol.* **44**, 641.
Bauer, W. D., Talmadge, K. W., Keegstra, K. and Albersheim, P. (1973). *Pl. Physiol.* **51**, 174.
Becker, G. E., Hui, P. A. and Albersheim, P. (1964). *Pl. Physiol.* **39**, 913.
Björndal, H., Hellerquist, C. G., Lindberg, B. and Svensson, S. (1970). *Angew. Chem. Int. Ed. Engl.* **9**, 610.
Bonner, J. (1934). *Protoplasma* **21**, 406.
Brauner, L. and Diemer, R. (1967). *Planta* **77**, 1.
Cleland, R. (1971). *A. Rev. Pl. Physiol.* **22**, 197.
English, P. D. and Albersheim, P. (1969). *Pl. Physiol.* **44**, 217.
English, P. D., Maglothin, A., Keegstra, K. and Albersheim, P. (1972). *Pl. Physiol.* **49**, 293.
Etherton, B. (1970). *Pl. Physiol.* **45**, 527.
Evans, M. L., Ray, Peter M. and Reinhold, L. (1971). *Pl. Physiol.* **47**, 335.
Ghuysen, J. M (1968) *Bact. Rev.* **32**, 425.
Grant, G. T., McNab, C., Rees, D. A. and Skerrett, R. J. (1969). *Chem. Commun.* p. 805.
Hager, A. (1969). *Planta* **89**, 224.
Hager, A., Menzel, H. and Krauss, A. (1971). *Planta* **100**, 47.
Hakomori, S. (1964). *J. Biochem.* (*Tokyo*) **55**, 205.
Jaffe, M. J. (1970). *Pl. Physiol.* **46**, 768.
Jagendorf, A. T. and Uribe, E. (1966). *Proc. natn. Acad. Sci. U.S.A.* **55**, 170.
Jones, T. M. and Albersheim, P. (1972). *Pl. Physiol.* **49**, 926.
Karlish, S. and Avron, M. (1968). *Biochim. biophys. Acta* **153**, 878.

Keegstra, K., Talmadge, K. W., Bauer, W. D. and Albersheim, P. (1973). *Pl. Physiol.* **51**, 188.

Kitasato, H. (1968). *J. gen. Physiol.* **52**, 60.

Kooiman, P. (1960). *Acta bot. Neerl.* **9**, 208.

Lamport, D. T. A. (1965). *Adv. bot. Res.* **2**, 151.

Lamport, D. T. A. (1970). *A. Rev. Pl. Physiol.* **21**, 235.

Lamport, D. T. A. and Miller, D. H. (1971), *Pl. Physiol.* **48**, 454.

Lamport, D. T. A., Katona, L. and Roerig, S. (1973) *Biochem. J.* **133**, 125.

Martin, D. L., Melancon, M. J., Jr. and DeLuca, H. F. (1969). *Biochem. biophys. Res. Commun.* **35**, 819.

Melancon, M. J., Jr. and DeLuca, H. F. (1970). *Biochemistry* **9**, 1658.

Mitchell, P. (1965). *Nature, Lond.* **208**, 147.

Nakazawa, T., Asami, K., Shoger, R., Fujiwara, A. and Yasumasu, I. (1970). *Expl. Cell Research* **63**, 143.

Nitsch, J. P. and Nitsch, C. (1956). *Pl. Physiol.* **31**, 94.

Rayle, D. L. and Cleland, R. (1970). *Pl. Physiol.* **46**, 250.

Rees, D. A. and Wight, N. J. (1969). *Biochem. J.* **115**, 431.

Rees, D. A. and Wight, A. W. (1971). *J. chem. Soc.* **(B)**, 1366.

Roelofsen, P. A. (1959). *In* "The Plant Cell Wall", p. 128. Borntraeger, Berlin.

Simmons, D. A. R. (1971). *Eur. J. Biochem.* **18**, 53.

Stowe, B. and Dotts, M. (1971). *Pl. Physiol.* **48**, 559.

Talmadge, K. W., Keegstra, K., Bauer, W. D. and Albersheim, P. (1973). *Pl. Physiol.* **51**, 158.

Timell, T. E. (1965). *Adv. Carbohyd. Chem.* **20**, 409.

Weigl, J. (1969). *Z. Naturforsch.* **24b**, 365.

Wilder, B. and Albersheim, P. (1973). *Pl. Physiol.* **51**, 889.

CHAPTER 11

Sites of Synthesis of the Polysaccharides of the Cell Wall

D. H. NORTHCOTE

Department of Biochemistry, University of Cambridge, England

I. INTRODUCTION

The cell wall is a constantly changing feature of the plant cell from the time of its inception as the cell plate at telophase to the time of its full development when it may be a completely lignified, massively thickened, secondary wall of a tracheid.

It has to accommodate the increase in size and the changing function of the cell during its development. Its structure contributes in a large number of cases to the function of the cell in the whole plant especially when the cell has a skeletal, conducting or protective role. In some cells which form a continuous channel the end wall that connects contiguous cells is completely removed or is modified to form a sieve perforated by wide pores. Similarly modified pores or connections, between laterally connected cells, which sometimes can be opened or closed are made up from modified wall material. Throughout its formation the wall has to respond to the living protoplast which makes it and which lays down the wall in a sequential developmental pattern.

The wall is a biological composite material in which a microfibrillar mesh is embedded within a continuous matrix (Mark, 1967; Northcote, 1972). The two phases interact to accommodate changes in size and shape and yet retain the mechanical and protective properties of the wall. The changing feature of the wall is its important structural component, water. This change is achieved by a variation in the composition and nature of the matrix material of the wall. The polymers of the matrix are hydrophilic and their conformations depend upon the presence of water, and since the interactions between the microfibrils and the matrix are related to the water content of the wall any change in the matrix polymers affects the mechanical properties of the wall (Northcote, 1972).

The synthetic processes produce a continual supply of cellulose microfibrils coincidentally with a changing supply of matrix material, finally terminated by a synthesis of hydrophobic lignin. The polysaccharide formation and deposition is achieved by the use of a membrane flow system when the polysaccharides are either synthesized within this system and then transported to the outside of the cell or the enzyme for the synthesis are conveyed to the outside of the plasmalemma. On the other hand the lignin can be polymerized and deposited within the wall from small molecular weight precursors supplied by other cells and when the extensive lignification takes place the cell is lignified from the outside inwards (Freudenberg, 1968).

II. THE MEMBRANE SYSTEM

The Golgi apparatus in conjunction with the endoplasmic reticulum, membrane-bounded vesicles and lysosomes forms an extensive intercommunicating membrane system within the cell. The lumen enclosed by this membrane system is made up by that of the perinuclear space, the endoplasmic reticulum, the cisternae, vesicles and tubules of the Golgi complex, the lysosomes, the pinocytotic and autophagic vesicles and the vacuole that is contained by the tonoplast. The region outside the plasmalemma is also included as part of this intercommunicating transport route. All these spaces are connected either by a functional continuity or by a direct morphological union so that together they form a distinct channel that is separated from the rest of the cytoplasm and nucleoplasm of the cell and in which material can be synthesized, modified and transported (Northcote, 1970; Dauwalder et al., 1972).

A. THE GOLGI APPARATUS

The Golgi apparatus occupies a key position in the functioning of this membrane system since not only are specific synthetic functions carried out

within the organelle and its associated vesicles but also, as part of the transport route, it presents an irreversible connection between the perinuclear
space and endoplasmic reticulum on the one hand and the vesicles and plasmalemma on the other. Because it is also a relatively localized region of the
diffuse membrane system it represents a site along the route where synthesis
and transport could be controlled (Northcote, 1970).

Many of the cytological, biochemical and chemical studies on the Golgi
apparatus and the endoplasmic reticulum can be interpreted by postulating
that there is a flow of membrane material from the endoplasmic reticulum
via the Golgi apparatus to the plasmalemma (Caro and Palade, 1964;
Jamieson and Palade, 1967a,b; 1968a,b; 1971a,b). The most likely theory
for the formation of the Golgi apparatus is that the cisternal membranes
arise from the rough endoplasmic reticulum which changes to smooth endoplasmic reticulum and which then becomes the Goldi cisternae and that
these cisternae break down to vesicles which can fuse with and extend the
plasmalemma. The maintenance of the Golgi complex requires the presence
of the nucleus (Flickinger, 1969; 1971). There are experimental results which
indicate that there are transitional elements of the endoplasmic reticulum
between the rough endoplasmic reticulum and the Golgi cisternae (Caro
and Palade, 1964; Jamieson and Palade, 1967a,b). The outermost cisternae
of the Golgi apparatus could be constantly formed by the fusion of vesicles
derived from the reticulum system while the inner cisternae in any one stack
could become fenestrated and could pinch off vesicles: this has given rise
to the idea that the Golgi apparatus may have a newly forming face at one
surface and a mature or secreting face at the opposite surface. For some
Golgi bodies the entire organelle is formed and reformed within 20–40 min
and individual cisternae are released once every 2–4 min (Neutra and Leblond,
1966; Brown, 1969).

Chemical analysis of the membranes of the Golgi apparatus has indicated
that they have a composition which is between that of the endoplasmic
reticulum and of the plasmalemma (Keenan and Morré, 1970). This has also
been suggested as a result of a study of the cytoplasmic membranes of the
plant pathogenic fungus *Pythium ultimum* Trow in which the overall membrane thickness, the staining intensity and the substructural pattern of stain
deposition were investigated. It was shown that at one extreme the plasmalemma and the membranes of the vesicles (free in the cytoplasm and attached
to Golgi cisternae at the mature face) stained intensely, were thicker (7·5 nm)
and were clearly unit membranes while at the other extreme the nuclear
envelope and the endoplasmic reticulum membranes stained faintly, appeared
thinner (52–40 Å) and rarely revealed the dark–light–dark pattern of the
unit membrane. The membranes of the Golgi apparatus were differentiated
across the stack so that those at the forming face appeared similar to those
of the endoplasmic reticulum and nucleus whereas those at the mature face
resembled the plasmalemma; and membranes of the mid-region of the stack

were intermediate in their appearance (Grove *et al.*, 1968). A similar differentiation between the membranes of the endoplasmic reticulum, Golgi and plasmalemma can be seen in higher plants using freeze-substitution methods and these appearances certainly reflect chemical differences between the membranes at the different sites (Hereward and Northcote, 1972). The results all indicate that the membranes are modified within the Golgi stack and that one of the functions of the apparatus is to alter the membranes of the endoplasmic reticulum so that they resemble that of the plasmalemma. Presumably this process is irreversible except by disaggregation of the membrane subunits and by their reassembly. Thus the Golgi apparatus represents a one-way valve within the system: nuclear envelope → endoplasmic reticulum → vesicles → Golgi apparatus → vesicles → plasmalemma.

III. Sites of Hemicellulose and Pectin Synthesis

A general feature of the membrane system is that it is concerned with the formation and packaging of material for export across the plasmalemma by a process of reverse pinocytosis. This particular function is of prime importance for the formation of the plant cell wall (Northcote, 1970; Dauwalder *et al.*, 1972).

Most of the early evidence for the function of the membrane system in cell wall production came from direct observation of the ultrastructure of dividing cells (formation of cell plate) (Hepler and Newcomb, 1967; Lehmann and Schulz, 1969; Esau and Gill, 1965; Roberts and Northcote, 1970; Whaley and Mollenhauer, 1963; Whaley *et al.*, 1966) root cap, root hairs and meristematic cells (formation of slime and primary cell walls) (Sievers, 1965; Schnepf, 1969; Mollenhauer and Whaley, 1963; Mollenhauer *et al.*, 1961) developing xylem and phloem (formation of secondary thickening and specialized pore and wall interconnections) (Wooding and Northcote, 1964; Pickett-Heaps and Northcote, 1966c; Northcote and Wooding, 1966). In all these observations the importance of the Golgi apparatus and the endoplasmic reticulum during the wall formation was indicated and incorporation of material into the cell wall by the reverse pinocytosis of material contained in single membrane-bounded vesicles was apparent.

Direct experimental evidence which clearly demonstrated a flow of material from the inside to the outside of the cell and a possible modification of its structure within the membrane system of the cell can be obtained by the use of radioactive markers. The radioactive tracer is supplied as a pulse to the cells so that it is incorporated into the material and is then chased through the system by a non-radioactive precursor. The passage of the labelled material through the cell can be followed by direct autoradiography of thin sections of the tissue prepared for electron microscopy. It can also be followed by determining the kinetics of the incorporation of the radioactivity supplied as a short pulse of the precursor into the tissue and by preparing from the tissue cell fractions which separate the various parts of the membrane system during the period of the pulse chase experiment.

By autoradiography it can be shown that radioactively labelled polysaccharide material appears inside the cisternae of the Golgi bodies of the cells of the root cap of wheat seedlings within five minutes of a period of incubation of the root with radioactively labelled glucose. This material is progressively transferred to the wall by incorporation into Golgi vesicles, passage across the cytoplasm, and eventual discharge into the wall by reverse pinocytosis. It can be isolated from the tissue and is found to consist of galactose, arabinose and galacturonic acid residues; it is therefore composed of polysaccharide(s) of the same type as the pectic substances (Northcote and Pickett-Heaps, 1966).

In the developing cells of xylem and phloem it has been shown by a similar autoradiographic study that xylans are deposited into the wall from the Golgi bodies (Northcote and Wooding, 1966; Wooding, 1968).

A. ISOLATED PARTS OF THE MEMBRANE SYSTEM

1. *Pea root*

A large percentage (*ca.* 63%) of the radioactivity present in the sugars of a hydrolysate of an isolated Golgi fraction obtained from pea-roots which had been incubated with uniformly labelled D-[^{14}C]glucose was in galactose, arabinose and galacturonic acid. These sugars are characteristic of the pectic substances. In addition the hydrolysate contained some radioactive xylose, mannose and a trace amount of radioactive glucuronic acid which indicated the occurrence of xylans and glucomannans (Harris and Northcote, 1971). The pattern of incorporation of radioactivity into the sugars, except for glucose, contained in the polymers of the isolated Golgi fraction was very similar to that incorporated into the polysaccharides of the wall during the incubation. Glucose was not labelled to a great extent relative to the sugars of the matrix polysaccharides and thus the Golgi apparatus is probably not active in the transport of cellulose.

2. *Maize root*

Much work has been done to characterize the polysaccharide components of maize roots (Roberts and Butt, 1967; 1969; Harris and Northcote, 1970; Kirby and Roberts, 1971) and it is possible to investigate changes in polysaccharide synthesis during differentiation of the tissues in the root and to show the involvement of different membrane fractions in the production of the polymers in particular zones of differentiation within the root (Bowles and Northcote, 1972).

D-[U^{14}C]Glucose was given to maize seedling roots *in vivo*. After 2 h incubation the roots were dissected into a root cap region (tip 1–2 mm) and a more mature region (at least 10 mm back from the tip). The tissues were homogenized in 0·5 M sucrose containing 0·1 M glutaraldehyde and the membrane fractions prepared by centrifugation on discontinuous gradients of sucrose solutions and the membrane fractions were caught on sucrose cushions of varying density. The membrane and other fractions that were isolated were identified as those rich in walls, Golgi bodies, mitochondria and smooth membranes and microsomes together with soluble polymers.

An analysis of the radioactivity of the sugars incorporated into the wall fraction indicated that the polysaccharide components present were dependent on the state of differentiation of the tissues. The values of the ratios of the radioactivity of galacturonic acid/glucuronic acid, galactose/glucose and arabinose/xylose were higher in wall regions of the root tip compared with values for wall region 20–30 mm from the tip. This suggests that these two regions of the root, the tip and the more mature region that were taken for preparation of the membrane fractions represented zones typical of primary and secondary wall development. In addition, the slime produced by maize roots contains fucose (32% of the neutral sugars), which is absent from any other polysaccharide that is typical of maize roots. Since slime production is restricted to the root-cap zone, and fucose is a marker for the presence of slime (Harris and Northcote, 1970), it follows that the maize root represents an ideal control system for the study of the involvement of different membrane fractions in production of slime from distinct root-cap and non-cap regions of the roots. The wall and slime can be regarded as polysaccharide constituents that have been synthesized within the cell-membrane system and then transported out of the cell.

TABLE I

Relative amounts of radioactivity incorporated from D-[U^{14}C]glucose into the polysaccharide components of the wall fraction from maize root tissue

Sugar	Radioactivity %		
	Whole Roots	Root cap tissue	Older tissue
Galacturonic acid	2·6	7·8	2·7
Glucuronic acid	1·2	1·3	0·8
Galactose	7·0	17·4	6·9
Glucose	63·8	38·3	62·5
Mannose	1·2	2·5	1·8
Arabinose	7·2	12·5	8·8
Xylose	15·7	13·5	15·6
Fucose	0·9	4·9	0·4
Ribose + Rhamnose	2·7	1·2	0·7
Total cpm	720 300	130 500	18 000

All the membrane fractions prepared from the two different zones of the root contained an appreciable amount of radioactivity incorporated into neutral sugars and sugar acids of polysaccharides. The percentage distribution of radioactivity in the polysaccharide components of the three membrane fractions was similar but the polysaccharides synthesized within the membrane compartments varied in the different zones of differentiation (see Tables I–III).

TABLE II

Relative amounts of radioactivity incorporated from D-[U-^{14}C]glucose into the polysaccharide components of the Golgi fraction from maize root tissue

Sugar	Radioactivity %		
	Whole roots	Root cap tissue	Older tissue
Galacturonic acid	14·0	11·6	10·3
Glucuronic acid	1·9	4·7	1·6
Galactose	26·6	34·1	21·5
Glucose	6·6	6·8	3·95
Mannose	3·5	1·8	1·0
Arabinose	22·3	14·7	19·8
Xylose	20·1	18·5	41·4
Fucose	2·8	6·2	0·2
Ribose + Rhamnose	1·7	1·7	—
Total cpm	4500	1100	790

TABLE III

Relative amounts of radioactivity incorporated from D-[U-^{14}C]glucose into the polysaccharide components of the microsome from maize root tissue

Sugar	Radioactivity %		
	Whole roots	Root cap tissue	Older tissue
Galacturonic acid	7·5	11·2	5·2
Glucuronic acid	1·7	1·6	1·2
Galactose	24·1	31·4	21·2
Glucose	4·0	7·3	8·5
Mannose	3·1	5·0	3·4
Arabinose	23·0	22·4	20·4
Xylose	31·0	16·4	38·8
Fucose	1·9	4·5	1·1
Total cpm	89 000	68 000	26 000

The lower radioactivity ratio of arabinose/xylose that occurred in the membrane fraction from the older region relative to the root tip, was attributable mainly to the high incorporation of glucose into the xylose component of polysaccharides in older root tissue membrane fractions. The arabinose/xylose ratio present in the wall is a measure of the relative amount of primary to secondary wall development (Thornber and North-cote, 1961) and this change is brought about by a corresponding change in the synthetic and transport function of the membrane system of the cell.

The percentage of radioactivity in galactose, arabinose and galacturonic acid was comparatively high in all the membrane fractions isolated from all the regions of the root which indicated that the synthesis of polysaccharides containing these monosaccharides (pectic substances) were maintained in these young roots.

The presence of radioactive fucose in the hydrolysate of the Golgi bodies and the other membrane fractions of the whole roots was direct evidence that the slime polysaccharide was produced by the membranes of the root-cap cells. When the membrane fractions from the root tip were examined the incorporation of glucose into fucose was greater than in the whole root, irrespective of the fact that much less root was used to provide the fractions. Membrane fractions from regions not containing root-cap tissue did not contain radioactive fucose to any significant extent.

The experiments show that any part of the membrane system at any one time carries polysaccharides (hemicelluloses, pectic substances or other soluble polymers) that are characteristic of the tissue from which the membranes were prepared. When the relative amounts of the membrane fractions isolated from a fixed amount of root tissue were measured in terms of the lipid content of the fraction it was shown that forty times as much microsomal membrane as Golgi fraction was present. If the relative specific radioactivity of the incorporation into polysaccharide material of each fraction was measured (cpm in polysaccharide per unit amount of lipid) then the Golgi fraction had at least twice the activity of that of the rough endoplasmic reticulum. The Golgi apparatus is thus an important localized focal point in the synthetic and transport system of polysaccharide. At the Golgi apparatus polysaccharide is concentrated in relation to other parts of the membrane system. This concentration at the Golgi could be caused by two factors. Radioactive sugar residues could be incorporated into growing polysaccharide chains as additional residues within the Golgi apparatus (as in the case of glycoprotein formation in animal tissues); also, polysaccharide synthesized within the diffuse membranes of the endoplasmic reticulum could undergo physical concentration since only a localized part of the membrane may be used to package and thus concentrate the polysaccharide for transport. The experiments also indicate that since all the membrane fractions that were isolated contained exportable polysaccharide, there may be a direct contribution from any part of the membrane system to the cell wall poly-saccharides.

IV. THE PLASMALEMMA AND THE SITE OF CELLULOSE SYNTHESIS

Although the wall fractions from all parts of the root contained a very high percentage of radioactivity incorporated into a glucan, most probably cellulose, the isolated parts of the membrane system did not contain radio-active glucans to any considerable extent (Harris and Northcote, 1971; Bowles and Northcote, 1972) (Tables I–III). This was in direct contrast to the pectic polysaccharides and hemicellulose. The radioactive polysaccharides found in the membrane fraction were contained within the membrane, and consequently indicated the function of those membranes in the synthesis and transport of polymers during the time of incubation with radioactive nutrient. Polysaccharide formed only at a membrane surface, and not contained within a compartment, would not be found in any of the fractions. The results, therefore, strongly suggest that even in regions of the plant where the greatest incorporation of radioactivity into the wall is into a glucan that is most probably cellulose, glucan is not contained within a membrane system, either during polymerization or for export to the wall. This is in accordance with the idea that cellulose is formed at the plasmalemma surface where the polymers are synthesized at specific loci and are organized into the basic network of microfibrils already present (Northcote, 1969b, 1972). Since the plasmalemma is formed from vesicle membranes and then maintained by the active addition of these membranes, it is possible that the enzyme complement necessary for β-$(1 \rightarrow 4)$-glucan production is present within the membrane system before its incorporation at the plasmalemma but that the activity remains latent until the environment at the cell surface is reached. Various studies have indicated that in some instances some β-$(1 \rightarrow 4)$-glucan can be synthesized within the membrane system possibly at the Golgi body stage. Brown et al. have investigated the polysaccharide scales found in certain flagellates (Brown, 1969; Brown et al., 1969; Herth et al., 1972; Brown et al., 1970). These are found within the Golgi complex before export across the plasmalemma and they can be seen to develop into their characteristic form within the Golgi apparatus and its associated vesicles so that some synthesis or assembly occurs in this part of the membrane system (Manton, 1966; 1967). The scales have dissimilar surfaces dorsoventrally and they are always oriented in a definite way within the vesicle. Since the scale continues to develop in this asymmetric way within the vesicle before its liberation to the outside of the cell, it would seem that the vesicle also has a dorsoventral asymmetry either in the lumen or more probably on the membranes. Brown and co-workers have shown that part of the polysaccharide complex that forms the scale is composed of glucose chains linked by β-$(1 \rightarrow 4)$-bonds and that it resembles cellulose in other chemical and physical properties.

A. THE PLASMALEMMA OF YEAST CELLS

The yeast cell wall does not contain cellulose but it does contain other polysaccharides, glucans and mannans, which make up about 60% of the wall (Northcote and Horne, 1952). Some of the wall material is formed off site in vesicles which are transported to and incorporated into the wall (Sentandreu and Northcote, 1969) and some appears to be formed at the plasmalemma surface (Northcote, 1969b). On the outer surface of the plasma-lemma there are regular hexagonal areas consisting of particles which are connected by fibres to the inner layer of the wall (Moor and Mühlethaler, 1963; Northcote, 1968a). These observations have been made on freeze-etched cells and indicate that wall material does arise from an organized part of the outer surface of the plasmalemma. Particles can be seen on the outer surface of the plasmalemma of higher plants and although these are not arranged in any organized pattern they can sometimes be observed to occur in rows (Northcote, 1968b; 1969a,b; Northcote and Lewis, 1968). Particles similar to those at the plasmalemma surface can also be found within the microfibrillar mass of the cell wall and from these particles microfibrils appear to radiate. This suggests again that the microfibrils may originate from particles at the outer membrane surface.

B. THE FORMATION OF THE PLASMALEMMA AND THE CELL PLATE AT CYTOKINESIS

Partially synchronized suspension cultures of sycamore cells afford an excellent system for studying the division and the accompanying fine structural organization during cytokinesis (Roberts and Northcote, 1970). The presence in the cells of an extensive vacuole is of especial interest during division since the cell plate is formed within a strand of cytoplasm which grows across the vacuole as the cell plate is extended. This strand can therefore be examined separately from the other more general cytoplasm of the cell and it contains all the organizational features which are necessary for cell plate formation. Frequently the nucleus is moved to the side of the cell at the mother cell wall just before division and at telophase the cell plate fuses with the mother cell wall at this side of the cell while it is still growing across the phragmosome to join the wall at the opposite side. Consequently, along the length of the cell plate from the mother cell wall back to the end of the phragmoplast it is possible to trace a complete sequence of events which shows the development of the plate as it grows to form a new cell wall. Within the cytoplasmic strand of the phragmoplast the organelles are arranged to form and extend the cell plate. Observations on this cytoplasm suggest that both the Golgi apparatus and the endoplasmic reticulum contribute to cell wall formation. At the growing tip of the cell plate vesicles, which have a characteristic appearance, are incorporated and these are almost certainly formed from the Golgi bodies found in this region; incorporation appears to occur into the matrix of the plate. The endoplasmic reticulum is found in a very recognizable form at both the cell plate and the growing wall and in some places it is found to have profiles which become closely applied to the plasmalemma. Fibrils, which resemble cellulose microfibrils, are visible in the centre of the new cell plate a short distance back from the growing tip so that these are woven into the assembled matrix. Further back in the plate at the more mature regions, the fibrils become progressively more abundant. Within the microfibrillar net fine particles can be seen and these suggest again that the α-cellulose microfibrils of the wall, unlike the matrix material, arise from a particulate system at the outer surface of the newly formed plasmalemma and that the plasmalemma was formed, at least in part, from the membranes of fused vesicles. These ideas also indicate that the fusion of the daughter cell wall into the mother cell wall is accomplished first by the extension of the matrix material as a ridge on the mother cell wall and then the microfibrils are woven into the matrix as the plate extends across the cytoplasm.

C. CELL WALL REGENERATION OF PROTOPLASTS

The sequence of events in wall formation can also be followed by studying the regeneration of the wall of plant protoplasts. Protoplasts prepared from soybean tissue culture cells regenerated an osmotically stable cell after 40 h culture (Hanke and Northcote, 1974). When the freshly prepared protoplasts were incubated with radioactive glucose it was shown that by 20 h pectin had been released to the medium and that the cells were still osmotically fragile. By 40 h the protoplast had formed a stable outer covering but nevertheless the wall at this time was devoid of pectin which was still being excreted into the medium and not retained at the outer surface of the cell. The wall consisted of glucan only and was presumably a network of microfibrils woven at the surface but not embedded in a matrix because although this was formed at an early stage of the incubation it was being continually lost to the medium. (This cannot happen during the formation of the matrix of the cell plate by vesicle fusion at cytokinesis.) The pectin excreted by protoplasts can be continuously isolated and analysed and the composition of these polysaccharides which are exported across the plasmalemma changes during the 40 h incubation. They are a form of weakly acidic pectinic acid in which a chain of polygalacturonic acids carries varying amounts of neutral arabinose and galactose units. These pectinic acids are assembled within the cell and excreted in a completed form.

D. CELLULOSE FORMATION IN THE WALL OF TRANSFER CELLS

During the formation of the extensive wall protuberances which are produced when transfer cells are induced in the roots of potato by infection with nematodes, the process of wall formation can also be seen (Jones and Northcote, 1972a,b). Here again there is evidence of the involvement of the plasma-lemma in the production of cellulose which is woven into a deposit of matrix material already present. At the local regions of protuberance formation the plasmalemma is extended and convoluted and microfibrillar material appears to arise from the outer surface of the membrane and is elaborated as a net into the matrix.

V. The Endoplasmic Reticulum

Although the amount of incorporation of radioactivity from glucose into polysaccharide present in an isolated rough endoplasmic reticulum fraction from the cells of maize roots is considerable, a direct contribution from the endoplasmic reticulum to the wall (not via the Golgi apparatus) has not yet been shown. Nevertheless the formation of wall material and the organiza-tion of material for transport into the wall is correlated in some instances with the appearance of endoplasmic reticulum profiles in a distinct pattern with respect to the developing wall and the sites of wall deposition. This is very apparent during cell plate formation and has already been described (p. 174).

A. XYLEM DEVELOPMENT

In a rapidly growing tissue such as the vascular regions of a wheat coleop-tile, profiles of the endoplasmic reticulum are found distributed in a regular manner between the thickenings of the walls of xylem vessels. These profiles represent sheets of endoplasmic reticulum within the strands of cytoplasm that are partly enclosed by the secondary thickenings that grow out from the primary wall into the cell. The profiles are arranged parallel to the thick-enings but more or less perpendicular to the primary wall (Pickett-Heaps and Northcote, 1966c).

B. PHLOEM DEVELOPMENT

A more complex relationship between the distribution of the endoplasmic reticulum and the form of the cell wall can be observed in the development of the sieve-tube of the phloem tissues. In these cells the sieve plate area is marked out by profiles of endoplasmic reticulum on each side of the wall (Esau et al., 1962; Northcote and Wooding, 1966). These profiles lie close to the plasmalemma and overlie plasmodesmata connecting the adjacent sieve-tube cells. As the sieve plate develops, normal wall material is deposited except at the sites of the profiles of endoplasmic reticulum around the plasmodesmata. At these positions, cones of callose (β-(1 $\rightarrow$ 3)-glucan) are formed. These cones meet with their apices at the centre of the wall and bulge out of the general line of the wall. The base of each cone is thus covered in each sieve tube cell by the plasmalemma and the profile of endoplasmic reticulum. The presence of the endoplasmic reticulum covering an area of the wall surface of the sieve plate results in the deposition of a particular polysaccharide. Another specialized and complex pore which connects the sieve-tube with the accompanying companion cell is also formed in the wall in close proximity to profiles of the endoplasmic reticulum. In this pore too callose is laid down around the pore on the side which communicates with the sieve tube (Wooding and Northcote, 1965; Northcote and Wooding, 1968).

In the mature sieve-tube the pore is surrounded by callose; a plasmalemma is present but there is no endoplasmic reticulum. Most of the organelles such as the Golgi bodies, mitochondria, nucleus and plastids are absent from the functional lumen of the mature sieve-tube cell. Nevertheless if a section of stem is fed radioactive glucose and the mature sieve-tube is examined, radioactive material is found incorporated into the callose lining of the pores of the sieve plate (Northcote and Wooding, 1966). It is possible, therefore, that the synthesis of callose by enzymes still present within the callose of the sieve plates or by enzymes at the plasmalemma or by enzymes in both situations can take place. The enzymes necessary for callose synthesis at these sites may have been provided by the endoplasmic reticulum profiles which marked out these regions at the initial stages of pore formation.

VI. GLYCOPROTEIN DEPOSITION IN THE WALL

In sycamore suspension cells both autoradiographical and chemical analysis have confirmed the idea that most of the protein that contains hydroxyproline is located within the cell wall (Lamport and Northcote, 1960; Roberts and Northcote, 1972). This protein is part of a glycoprotein and it carries short oligosaccharide chains (approx. 9 residues) containing arabinose and galactose residues (Heath and Northcote, 1971, 1973). These chains are attached via arabinose to the hydroxyproline units (Lamport, 1967). Since the cells can be partially synchronized it is possible to use radioactive proline to investigate the time of incorporation of the glycoprotein into the wall and also to indicate the probable mechanism of its deposition. The studies showed that the glycoprotein was laid down both at division and in the growing cell (Roberts and Northcote, 1972). The more mature cells which were not dividing incorporated relatively more of the protein than the actively dividing cells. A little of the material was deposited in the forming cell plate but more was laid down in the growing wall and here it was deposited throughout the matrix. No evidence was obtained that the glycoprotein was exported within a vesicle derived from the Golgi apparatus, and if it is moved to the wall from the membrane system then it seems more likely that the endoplasmic reticulum part of the system is involved (Roberts and Northcote, 1972).

VII. Vesicle Transport and Membrane Fusion

Movement of vesicles carrying polysaccharide and their incorporation into the wall or their fusion within the mitotic spindle is a characteristic mechanism for wall or cell plate formation. The process, in addition to polysaccharide synthesis and packaging, necessitates an organized transport of the vesicles to definite positions within the cell and also a method for membrane fusion (Northcote, 1971).

The mechanism for the movement of the vesicles is unknown and seems to be part of the general cytoplasmic streaming which can be seen in normal, active living cells. The direction of movement is in part controlled by the distribution of microtubules (Northcote, 1969c, 1971; Roberts and Northcote, 1970). During the construction of the cell plate vesicles can be seen to be organized in rows between the microtubules and these rows are directed towards the cell plate region (Pickett-Heaps and Northcote, 1966a,b). In xylem the secondary thickening of the wall is usually laid down in a reticulate or spiral form on the primary wall. The microtubules are restricted to the regions of secondary thickening and are located just under the plasmalemma running around the periphery of the cell (Wooding and Northcote, 1964; Pickett-Heaps and Northcote, 1966c).

The orientation of the microtubules in the cytoplasm is often found to be the same as that of the microfibrils of cellulose in the wall but the connection between the two is not clear (Wooding and Northcote, 1964). The parallel orientation may be partly due to the flow into the wall of material which is directed by the microtubules and hence imposes a direction on the developing microfibrils (Northcote, 1969a,b).

The plasmalemma enclosing the cell plate of plants arises from the membranes of the vesicles which fuse together to form the plate and some of these vesicles are formed from the Golgi apparatus so that at this time there is a direct contribution of the membrane developed at the Golgi complex to the new plasmalemma that is formed at telophase (Roberts and Northcote, 1970). The Golgi apparatus continues to contribute to the extension of the plasmalemma during the growth in area of the cell wall when material contained within membrane-bounded vesicles that are derived from the Golgi apparatus is passed into the wall by fusion of the vesicles with the cell membrane (Pickett-Heaps and Northcote, 1966b). During secondary thickening of the wall there is still an extensive incorporation of material from the Golgi apparatus into the wall (Wooding and Northcote, 1964; Northcote and Wooding, 1966, 1968) and a corresponding contribution to the plasmalemma without any very great increase in surface area of the cell membrane. It is possible, therefore, that membrane material can be continuously formed and aggregated at the endoplasmic reticulum and Golgi apparatus and possibly disaggregated and released back to the cytoplasm at the plasmalemma.

The change and modification of membrane material which is thought to occur at the Golgi complex ensures that material packed in membranes and pinched off from the secretory face of the Golgi apparatus is surrounded by membrane of a similar nature to that of the plasmalemma. Hence, reverse pinocytosis at the outer cell membrane is more possible than fusion with the endoplasmic reticulum or other internal cell membrane.

Model systems for membrane fusion have been elaborated by a detailed study of the ultrastructure of the fusion of vesicles with the plasmalemma of vascular endothelia (Palade and Bruns, 1968), mast cell secretion (Lagunoff, 1973) and of mucocyst secretion in *Tetrahymena* (Satir *et al.*, 1973). The mucocyst is a membrane-bounded secretory vesicle and it moves to a distinct site just under the plasma membrane. By means of the freeze-etch technique this site can be identified as a rosette composed of particles (8–11 particles, each 150 Å diameter) which can be clearly seen at the surface. An annulus of particles (5–7 rows of closely packed particles, each 110 Å diameter) forms on the mucocyst membrane immediately underneath the rosette. The membranes fuse by a depression which grows within the plasma membrane rosette and as the depression increases it enlarges from 60 nm to 200 nm diameter. At the extreme tip of the mucocyst there is a circular piece of membrane (61 nm diameter) which is devoid of particles and it is likely that the rosette fits over this. A mechanism for membrane fusion can be described by interpreting the appearance of the annulus particles in the images of the freeze-fractured plasma membrane. These become visible in the surface views of the plasma membrane at a definite stage and if the freeze-fracture splits the membrane at the hydrophobic interface and the particles are seen in this region then a model for the fusion can be made.

Apart from the information about the events which occur at fusion the observations clearly indicate that the plasma membrane can be marked for vesicle fusion so that sites for fusion can be temporarily indicated at definite places at the cell surface.

REFERENCES

Bowles, D. J. and Northcote, D. H. (1972). *Biochem. J.* **130**, 1133.
Brown, R. M. (1969). *J. Cell Biol.* **41**, 109.
Brown, R. M., Franke, W. W., Kleinig, H., Falk, H. and Sitte, P. (1969). *Science, N.Y.* **166**, 894.
Brown, R. M., Franke, W. W., Kleinig, H., Falk, H. and Sitte, P. (1970). *J. Cell Biol.* **45**, 246.
Caro, L. G. and Palade, G. E. (1964). *J. Cell Biol.* **20**, 473.
Dauwalder, M., Whaley, W. G. and Kephart, J. E. (1972). *Sub-cell. Biochem.* **1**, 225.

Esau, K. and Gill, R. H. (1965). *Planta* **67**, 168.
Esau, K., Cheadle, V. I. and Risley, E. B. (1962). *Bot. Gaz.* **123**, 233.
Flickinger, C. J. (1969). *Anat. Rec.* **163**, 39.
Flickinger, C. J. (1971). *J. Cell Biol.* **49**, 221.
Freudenberg, K. (1968). *In* "Constitution and Biosynthesis of Lignin", p. 45. Molecular Biology, Biochemistry and Biophysics **2** (A. Kleinzeller *et al.*, eds), Springer-Verlag, Berlin.
Grove, S. N., Bracker, C. E. and Morré, D. J. (1968). *Science, N.Y.* **161**, 171.
Hanke, D. and Northcote, D. H. (1974). *J. Cell Sci.* **14**, 29.
Harris, P. J. and Northcote, D. H. (1970). *Biochem. J.* **120**, 479.
Harris, P. J. and Northcote, D. H. (1971). *Biochim. biophys. Acta* **237**, 56.
Heath, M. F. and Northcote, D. H. (1971). *Biochem. J.* **125**, 953.
Heath, M. F. and Northcote, D. H. (1973). *Biochem. J.* **135**, 327.
Hepler, P. K. and Newcomb, E. H. (1967). *J. Ultrastruct. Res.* **19**, 498.
Hereward, F. V. and Northcote, D. H. (1972). *Exp. Cell Res.* **70**, 73.
Herth, W., Franke, W. W., Stadler, J., Bettiger, H., Keilich, G. and Brown, R. M. (1972). *Planta* **105**, 79.
Jamieson, J. D. and Palade, G. E. (1967a). *J. Cell Biol.* **34**, 577.
Jamieson, J. D. and Palade, G. E. (1967b). *J. Cell Biol.* **34**, 597.
Jamieson, J. D. and Palade, G. E. (1968a). *J. Cell Biol.* **39**, 580.
Jamieson, J. D. and Palade, G. E. (1968b). *J. Cell Biol.* **39**, 589.
Jamieson, J. D. and Palade, G. E. (1971a). *J. Cell Biol.* **48**, 503.
Jamieson, J. D. and Palade, G. E. (1971b). *J. Cell Biol.* **50**, 135.
Jones, M. G. K. and Northcote, D. H. (1972a). *J. Cell Sci.* **10**, 789.
Jones, M. G. K. and Northcote, D. H. (1972b). *Protoplasma* **75**, 381.
Keenan, T. W. and Morré, D. J. (1970). *Biochemistry* **9**, 19.
Kirby, K. G. and Roberts, R. M. (1971). *Planta* **99**, 211.
Lagunoff, D. (1973). *J. Cell Biol.* **57**, 252.
Lamport, D. T. A. (1967). *Nature, Lond.* **216**, 1322.
Lamport, D. T. A. and Northcote, D. H. (1960). *Nature, Lond.* **188**, 685.
Lehmann, H. and Schulz, D. (1969). *Planta* **85**, 313.
Manton, I. (1966). *J. Cell Sci.* **1**, 375.
Manton, I. (1967). *J. Cell Sci.* **2**, 411.
Mark, R. E. (1967). "Cell Wall Mechanics of Tracheids." Yale University Press, New Haven, London.
Mollenhauer, H. H. and Whaley, W. G. (1963). *J. Cell Biol.* **17**, 222.
Mollenhauer, H. H., Whaley, W. G. and Leech, H. J. (1961). *J. Ultrastruct. Res.* **5**, 193.
Moor, H. and Mühlethaler, K. (1963). *J. Cell Biol.* **17**, 609.
Neutra, M. and Leblond, C. P. (1966). *J. Cell Biol.* **30**, 119.
Northcote, D. H. (1968a). *Br. med. Bull.* **24**, 107.
Northcote, D. H. (1968b). *In* "Plant cell organelles" (J. B. Pridham, ed.), p. 179. Academic Press, London and New York.
Northcote, D. H. (1969a). *In* "Essays in Biochemistry" (P. N. Campbell and G. D. Greville, eds), Vol. 5, p. 89. Academic Press, London and New York.
Northcote, D. H. (1969b). *Proc. R. Soc.* B **173**, 21.
Northcote, D. H. (1969c). *Symp. Soc. gen. Microbiol.* **19**, 333.
Northcote, D. H. (1970). *Endeavour* **30**, 26.
Northcote, D. H. (1971). *Symp. Soc. exp. Biol.* **25**, 51.
Northcote, D. H. (1972). *A. Rev. Pl. Physiol.* **23**, 113.
Northcote, D. H. and Horne, R. W. (1952). *Biochem. J.* **51**, 232.

Northcote, D. H. and Lewis, D. R. (1968). *J. Cell Sci.* **3**, 199.
Northcote, D. H. and Pickett-Heaps, J. D. (1966). *Biochem. J.* **98**, 159.
Northcote, D. H. and Wooding, F. B. P. (1966). *Proc. R. Soc.* B **163**, 524.
Northcote, D. H. and Wooding, F. B. P. (1968). *Sci. Prog.*, Oxford **56**, 35.
Palade, G. E. and Bruns, R. R. (1968). *J. Cell Biol.* **37**, 633.
Pickett-Heaps, J. D. and Northcote, D. H. (1966a). *J. Cell Sci.* **1**, 109.
Pickett-Heaps, J. D. and Northcote, D. H. (1966b). *J. Cell Sci.* **1**, 121.
Pickett-Heaps, J. D. and Northcote, D. H. (1966c). *J. exp. Bot.* **17**, 20.
Roberts, R. M. and Butt, V. S. (1967). *Expl Cell Res.* **46**, 495.
Roberts, R. M. and Butt, V. S. (1969). *Planta* **84**, 250.
Roberts, K. and Northcote, D. H. (1970). *J. Cell Sci.* **6**, 299.
Roberts, K. and Northcote, D. H. (1972). *Planta* **107**, 43.
Satir, B., Schooley, C. and Satir, P. (1973). *J. Cell Biol.* **56**, 153.
Schnepf, E. (1969). *Protoplasmatologia* **8**, 1.
Sentandreu, R. and Northcote, D. H. (1969). *J. gen. Microbiol.* **55**, 393.
Sievers, A. (1965). *In* "Funktionelle und morphologische Organisation der Zelle: Sekretion und Exkretion Funktion des Golgi-Apparates in pflanzlichen und tierischen Zellen", p. 89. Springer, Berlin.
Thornber, J. P. and Northcote, D. H. (1961). *Biochem. J.* **81**, 455.
Whaley, W. G., Dauwalder, M. and Kephart, J. E. (1966). *J. Ultrastruct. Res.* **15**, 169.
Whaley, W. G. and Mollenhauer, H. H. (1963). *J. Cell Biol.* **17**, 216.
Wooding, F. B. P. (1968). *J. Cell Sci.* **3**, 71.
Wooding, F. B. P. and Northcote, D. H. (1964). *J. Cell Biol.* **23**, 327.
Wooding, F. B. P. and Northcote, D. H. (1965). *J. Cell Biol.* **24**, 117.

CHAPTER 12

The Relation of Plant Enzyme-catalysed β-(1,4)-Glucan Synthesis to Cellulose Biosynthesis *in vivo**

C. L. VILLEMEZ

Division of Biochemistry and Department of Chemistry
University of Wyoming, Laramie, Wyoming, U.S.A.

I. INTRODUCTION

The higher plant cell wall is one of the most intricate of biological structures. Being closely involved in all stages of the growth and development
of essentially every plant cell, the cell wall is a structure that is in the process
of constant change. Yet, the trend of recent evidence implies that almost
every aspect of the wall structure, from the composition and sequence of
the monosaccharide constituents of the individual polysaccharides to the
interactions between the component macromolecules, is genetically determined. To achieve this level of control over the composition and spatial
arrangement of such a complex array of macromolecules requires an equally
complex and well regulated series of biosynthetic processes. A description
of the involvement of certain subcellular organelles in cell wall formation
can be found in Chapter 11 by Northcote in this volume.

Recently, basic information concerning the structure of the primary cell
wall was presented in a series of three papers from the laboratory of Peter
Albersheim (Talmadge *et al.*, 1973; Bauer *et al.*, 1973; and Keegstra *et al.*,
1973; see also Chapter 10 in this volume). On the basis of this evidence, there
appears to be a highly specific, non-covalent interaction between the covalently
linked non-cellulosic portion of the wall and the cellulose microfibrils. This
interaction is such that the directional orientation of the cellulose microfibrils

* Wyoming Experiment Station Journal Article No. JA 592.

appears to provide the pattern which determines the superstructure of the cell wall. Therefore, the processes which serve to orient the cellulose fibrils would appear to be one of the more significant determinants of cell wall structure. Unfortunately, very little is known about this aspect of cellulose biosynthesis. (For more information about this and other aspects of cellulose structure, see the review by Shafizadeh and McGinnis, 1971.) Another important biosynthetic problem relative to cellulose formation arises from the fact that the cellulose molecules found in the higher plant secondary wall appear to be not only of very large molecular weight, but the molecular weight range of these molecules is essentially zero. This information arises mainly from the studies of Marx-Figini (see, for example, Marx-Figini, 1966 and 1971). The biosynthetic consequences of this situation is that not only must more than 10 000 D-glucose molecules be joined to form a linear macromolecule, but the polymerization reactions must be terminated in a very exact manner. Among biological polymerization reactions, only the nucleic acids and proteins are synthesized in a manner which produces a monodisperse product. No organic chemical procedures presently known are capable of producing monodisperse polymers of such high degree of polymerization. Nucleic acids and proteins are synthesized using a template and it is this aspect of the synthetic process which produces products of monodisperse molecular weight. It seems unlikely that cellulose is synthesized using a coding process as complex as that for the nucleic acids and proteins. Some sort of template would appear to be the most likely factor which causes the specific termination of the polymerization reaction. However, there is no information presently available which would allow one to conclude that a template is actually involved in cellulose biosynthesis. Such information will most likely come from investigations of the properties of the enzyme system which catalyzes the formation of cellulose. This report is mainly concerned with the identification of that particular biosynthetic enzyme system.

II. PROPERTIES OF POLYSACCHARIDE SYNTHESIZING ENZYME SYSTEMS

A large number of enzyme systems which catalyze the polymerization of sugars have been studied. There are numerous reviews of the subject, one of the more recent of which is that by Nikaido and Hassid (1971). This section will consider only a very few properties of plant polysaccharide synthesizing enzymes which appear particularly relevant to the interpretation of results obtained using these systems.

A recent report (Panayotatos and Villemez, 1973) of the synthesis *in vitro* of a β-(1,4)-linked D-galactan chain contains two observations of relevance to interpreting the results of *in vivo* plant polysaccharide biosynthesis. Firstly, the [^{14}C]-β-(1,4)-linked D-galactan was insoluble in hot alkali. Alkali insolubility is a property usually associated with cellulose. To my knowledge, no alkali insoluble β-(1,4)-linked galactans have been isolated from plant sources.

Secondly, no alkali insoluble galactose-containing molecules could be found in the cell walls of *Phaseolus aureus* seedlings, the plant source for the *in vitro* enzyme system used in these studies. Since the enzyme system was quite active, very specific with regard to substrate utilized (UDP-D-galactose) and product synthesized, it seems likely that the system indeed serves the function of synthesizing a β-(1,4)-linked galactan chain. The reason for the alkali insolubility of the synthetic product may lie in two areas: (1) the natural product could contain side chains which would limit the association of the galactan with the powdered cellulose added as carrier; and (2) no data are available, to my knowledge, concerning the conditions necessary to remove by alkali extraction small amounts of short chain polysaccharide bonded to large quantities of insoluble cellulose. It would appear that considerable caution is required in using solubility data to relate polysaccharides synthesized with *in vitro* enzyme systems to the naturally occurring wall polysaccharides. A second observation, of relevance here, was derived from the same work. The galactose moieties which made up the tri- and tetra-saccharides obtained from acetolysis of the original [¹⁴C]galactan, were unequally labeled with the labeling heaviest at the reducing end of these oligosaccharides. In contrast, the disaccharide was uniformly labeled. These results imply two things: (1) only approximately two [¹⁴C]galactose moieties were incorporated into the galactan chain, and (2) these were located at the *reducing* end of the polysaccharide chain. Since transglycosylation reactions, with sugar nucleotides as the glycosyl donor, necessarily result in glycosyl additions at the *non-reducing* end of a saccharide chain, the latter steps resulting in the synthesis of this polysaccharide do not involve sugar nucleotides as glycosyl donors. Also, since only a few radioactive sugar moieties are incorporated into the finished polysaccharide, structural information obtained from *in vitro* synthesized plant wall polysaccharides relates to only a limited area of the entire molecule. Other data indicating unequal labeling of a polysaccharide synthesized *in vitro* were reported in connection with studies on glucomannan biosynthesis (Heller and Villemez, 1972). Data of this type indicate that the radioactive sugar nucleotides used as substrates for *in vitro* polysaccharide-synthesizing systems contribute only a small proportion of the sugar residues which constitute the [¹⁴C]polysaccharide product. The remainder of the sugar residues are present in the enzyme preparation probably as non-nucleotide saccharide intermediates.

Another property of these enzyme systems which should be noted is that very small differences in enzyme preparation can cause considerable changes in the properties of the individual glycosyl transferases. We have noted shifts in pH optima of around two pH units, large changes in kinetics, and alterations in metal cofactor requirements, as a function of seemingly minor alterations in enzyme preparation (A. F. Clark and Villemez, C. L., unpublished—for detailed information see A. F. Clark, Doctoral Dissertation, Department of Chemistry, Ohio University, 1972). This circumstance may be responsible for the difficulty encountered in comparing results between individual research groups.

III. SYNTHESIS OF β-(1,4)-GLUCANS *in vitro*

Using higher plant enzymes, polysaccharides which contain β-(1,4)-linked D-glucose residues have been synthesized from two sugar nucleotides, UDP-D-glucose and GDP-D-glucose (see Nikaido and Hassid, 1971). With the alkali insoluble [^{14}C]polysaccharides formed from UDP-D-glucose, acetolysis or partial acid hydrolysis produced glucose oligosaccharides linked β-(1,3) as well as glucose oligosaccharides linked β-(1,4). Indirect data indicated that the β-(1,3)-linked oligosaccharides were derived from a different poly-saccharide, [^{14}C]callose (Villemez *et al.*, 1967; Ordin and Hall, 1968). The biosynthesis of callose from UDP-D-glucose using similar enzymes had been described earlier (Feingold *et al.*, 1958). Recently, what appeared to be a pure alkali insoluble β-(1,4)-linked glucan was produced using UDP-D-glucose as substrate (Clark and Villemez, 1972). This UDP-D-glucose utilizing enzyme system could possibly be the one responsible for the synthesis of cellulose *in vivo*.

In 1964, Barber *et al.* concluded that the substrate for cellulose biosynthesis in higher plants was GDP-D-glucose. On the basis of the data available at the time, this appeared to be the only logical conclusion. However, more recent data concerning the properties of this enzyme system indicate that the function of the GDP-D-glucose utilizing enzyme *in vivo* is as a component in the synthesis of a β-(1,4)-linked glucomannan (Villemez, 1971; Heller and Villemez, 1972).

Almost no information is available concerning the molecular weight of the alkali insoluble [^{14}C]glucose-containing polysaccharides formed from sugar nucleotides using plant enzymes. The difficulty involved in obtaining this type of data relates to the insolubility of the products and the fact that the only method of detection is via radioactive labeling. Only a small amount of the biosynthetic product is available and this is in the presence of relatively huge quantities of non-labeled polysaccharide material. To circumvent these difficulties, I have examined the molecular size of acetate derivatives of the radioactive polysaccharides. These derivatives were prepared, using 2 mg of cellulose powder as a carrier, by a scaled-down version of the method described by Tanghe *et al.* (1963). The precipitation step was not included, so the reaction mixture was diluted with an equal volume of methanol-methylene chloride (1:1, v/v) and applied immediately to a column ($1 \cdot 2 \times 23$ cm) of glass beads containing uniform pores of either 75 Å or 2000 Å. (The porous glass beads were obtained from Sigma Chemical Co. (St. Louis, Mo., U.S.A.).) The eluting solvent was methanol-methylene chloride (1:1, v/v) and 1 ml fractions were taken after the sample (1 ml volume) had permeated the column. The radioactive polysaccharides were prepared as described previously (Clark and Villemez, 1972; Heller and Villemez, 1972) using particulate enzymes from *Phaseolus aureus* seedlings.

From the data in Fig. 1, it is apparent that all of the [^{14}C]polysaccharide derived from GDP-[^{14}C]-glucose permeates the 75 Å pores, which excludes linear dextrans of 28 000 molecular weight. In fact, a considerable portion of this [^{14}C]polysaccharide completely permeates the 75 Å pores suggesting a molecular weight of less than 28 000 for this portion. Although the column has not been calibrated, the peak degree of polymerization for the [^{14}C] polysaccharide from GDP-[^{14}C]glucose can be estimated as about 40. Also

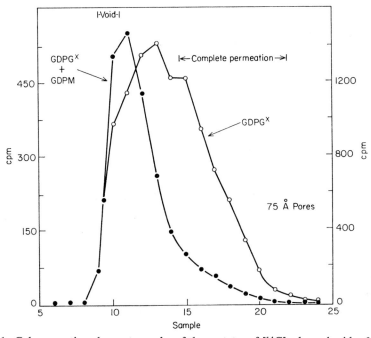

FIG. 1. Gel permeation chromatography of the acetates of [^{14}C]polysaccharides formed from GDP-D-[^{14}C]glucose. Details of the experimentation can be found in the text. Right hand vertical axis, solid circles; left hand vertical axis, open circles. 1 ml samples were collected.

in Fig. 1, the effect of adding unlabeled GDP-mannose on the alkali insoluble [^{14}C]polysaccharide from GDP-[^{14}C]glucose is to increase the peak degree of polymerization to about 80 and to eliminate the low molecular weight material. Timell (1965) reports that glucomannans obtained from angiosperms have an average degree of polymerization of 25–70 residues. Therefore, the molecular weight increase and the final molecular weight is consistent with the idea that the [^{14}C]polysaccharide material produced from GDP-[^{14}C] glucose alone consists of unfinished precursors of the final product and for completion of synthesis GDP-mannose is required. A comparison of the

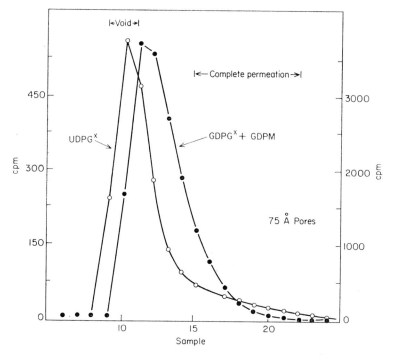

FIG. 2. Gel permeation chromatography of acetylated [¹⁴C]glucomannan and [¹⁴C] glucan. Details of the experimentation can be found in the text. Right hand vertical axis, solid circles; left hand vertical axis, open circles. 1 ml samples were collected.

elution pattern, with a 75 Å pore column, of the [¹⁴C]glucomannan and the [¹⁴C]polysaccharide produced from UDP-[¹⁴C]glucose is given in Fig. 2. The [¹⁴C]polysaccharide from UDP-[¹⁴C]glucose does not permeate the 75 Å pores to any significant extent, indicating that the molecular weight of the bulk of this polysaccharide is greater than 28 000. In Fig. 3, the data from a 2000 Å pore column indicate that a small proportion (about 7·5%) of the [¹⁴C]polysaccharide from UDP-[¹⁴C]glucose does not penetrate the pores at all. This indicates that the molecular weight of this fraction is greater than 1·2 million. From the [¹⁴C]glucomannan, there are no molecules in this size range. Since a molecular weight of this magnitude in cell wall polymers is found only in cellulose it seems likely that the enzyme system utilizing UDP-D-glucose to produce β-(1,4)-glucans indeed has a role *in vivo* in the synthesis of cellulose. The lower molecular weight material from UDP-[¹⁴C]glucose may be carbohydrate precursors of cellulose. These results are preliminary, and more details are necessary for firm conclusions. However, they suggest very strongly that studies of cellulose biosynthesis may profitably be concentrated on this particulate enzyme system utilizing UDP-D-glucose as a substrate.

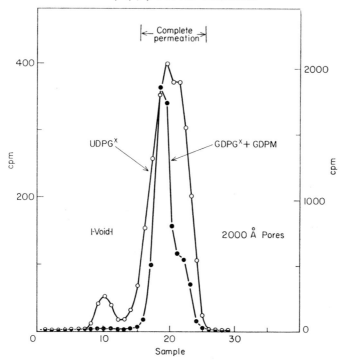

FIG. 3. Separation of high and low molecular weight acetylated [¹⁴C]glucan. Details of the experimentation can be found in the text. Right hand vertical axis, solid circles; left hand vertical axis, open circles. 1 ml samples were collected.

REFERENCES

Barber, G. A., Elbein, A. D. and Hassid, W. Z. (1964). *J. biol. Chem.* **239**, 4056.

Bauer, W. D., Talmadge, K. W., Keegstra, K. and Albersheim, P. (1973). *Pl. Physiol.* **51**, 174.

Clark, A. F. and Villemez, C. L. (1972). *Pl. Physiol.* **50**, 371.

Feingold, D. S., Neufeld, E. F. and Hassid, W. Z. (1958). *J. biol. Chem.* **233**, 783.

Heller, J. S. and Villemez, C. L. (1972). *Biochem. J.* **129**, 645.

Keegstra, K., Talmadge, K. W., Bauer, W. D. and Albersheim, P. (1973). *Pl. Physiol.*, **51**, 188.

Marx-Figini, M. (1966). *Nature, Lond.* **210**, 754 and 755.

Marx-Figini, M. (1971). *Biochim. biophys. Acta* **237**, 75.

Nikaido, H. and Hassid, W. Z. (1971). *Adv. Carbohydrate Chem.* **26**, 351.

Ordin, L. and Hall, M. A. (1968). *Pl. Physiol.* **43**, 473.

Panayotatos, N. and Villemez, C. L. (1973). *Biochem. J.* **133**, 263.

Shafizadeh, F. and McGinnis, G. D. (1971). *Adv. Carbohyd. Chem.* **26**, 297.

Talmadge, K. W., Keegstra, K., Bauer, W. D. and Albersheim, P. (1973). *Pl. Physiol.* **51**, 158.

Tanghe, L. J., Genung, L. B. and Mench, J. W. (1963). *Methods in Carbohydrate Chem.* **III**, 193.

Timell, T. E. (1965). *Methods in Carbohydrate Chem.* **V**, 137.

Villemez, C. L. (1971). *Biochem. J.* **121**, 151.

Villemez, C. L., Franz, G. and Hassid, W. Z. (1967). *Pl. Physiol.* **42**, 1219.

CHAPTER 13

Biosynthesis of Pectin and Hemicelluloses

HEINRICH KAUSS

Fachbereich Biologie der Universität, Kaiserslautern, West Germany

I. INTRODUCTION

For many decades the cell wall of higher plants was considered to be mainly composed of a mixture of well defined polysaccharide molecules held together by non-covalent forces. However, such a model never appeared satisfactory to plant physiologists. It did not serve as a satisfactory basis for understanding the striking properties of the wall, revealed by growth experiments (Cleland, 1971) or the relationship between ultrastructural observations and biogenesis and differentiation (see Northcote, this volume Chapter 11). At the present time there is growing evidence (Albersheim, this volume Chapter 10; Bolker and Wang, 1969; Lamport, 1970) that the different polysaccharide types can be interlinked by covalent bonds as well as bonded to the cell wall glycoprotein. However, we do not yet understand the exact molecular construction of the plant cell wall and, unfortunately, still cannot describe the growth mechanisms in terms of macromolecular biochemistry.

Nevertheless, we possess a considerable amount of information regarding the chemical structure of most of the polysaccharide building blocks of the plant cell wall and, consequently, we can try to examine the mechanisms for their biosynthesis. Villemez in this volume (see Chapter 12) gives a detailed survey of problems related to cellulose biosynthesis including some general pitfalls of which one should be aware when studying the synthesis of plant cell wall polysaccharides *in vitro*.

As some reviews dealing with the biosynthesis of cell wall polysaccharides have recently been published (Lamport, 1970; Kauss, 1970; Nikaido and Hassid, 1971; Northcote, 1972) this contribution will concentrate only on a few aspects of the biosynthesis of matrix polysaccharides.

II. SOURCE OF UDP-GLUCURONIC ACID

The oxidation of UDP-glucose to UDP-glucuronic acid is an important step in the Neufeld-Hassid pathway as this latter nucleotide in turn leads to numerous UDP-sugars which are necessary as donors for the biosynthesis of pectin and hemicelluloses. Loewus (1965) and co-workers have concluded from feeding experiments with myo-inositol-^{14}C that a pathway alternative to the one above leads to the production of free glucuronic acid by oxidative cleavage of the cyclitol and that this is followed by conversion of the uronic acid to UDP-glucuronic acid. As the enzymes required for these reaction, occur in plant tissues it was thought that the physiological significance of this pathway could possibly be to by-pass the formation of UDP-glucose and hence bring about some regulation at the cross-point of metabolism between glucans, such as cellulose, and matrix polysaccharides (Lamport 1970). In addition, it was assumed that in a similar pathway the methylated uronic acids could be derived from the naturally occurring inositol-methyl-ethers (Loewus, 1965). Feeding experiments, however, did not support the latter hypothesis (Loewus, 1971).

The data available on myo-inositol-^{14}C feeding experiments which led to the formulation of the above by-pass for the formation of UDP-glucuronic acid, does not generally allow one to estimate quantitatively its contribution to cell wall biosynthesis since the endogenous synthesis and pool size can never be taken into consideration. However, a recent study by Jung *et al.* (1972) with a strictly myo-inositol-dependent tissue culture of *Fraxinus pennsylvanica* shows that only 0·5% of the total galacturonic acid in the cell walls is derived from myo-inositol. The by-pass pathway, therefore, seems to be of minor importance, at least in this tissue, for the biosynthesis of cell wall material. The bulk of the cell wall carbon material appears to flow through UDP-glucose directly into the UDP-glucuronic acid pool.

The physiological role of the myo-inositol pathway outlined above appears more likely to be related to the degradation of myo-inositol which is an ubiquitous constituent of plant cells. The label from any surplus of myo-inositol-^{14}C fed into the cells thus ends up in glucuronic acid and necessarily flows into the pools used for biosynthesis of cell wall material, especially when cells or tissues in a state of extensive cell wall formation are used for such experiments.

III. BIOSYNTHESIS OF PECTIN

A. FORMATION OF THE MAIN CHAINS

There is now general agreement among all authors that the donors of monomeric sugar units for the biosynthesis of cell wall polysaccharides are nucleoside diphosphate sugars. In the case of the pectic substances the most prominent monomer is galacturonic acid, which forms the α-(1, 4)-linked backbone of this acidic polysaccharide. UDP-galacturonic acid can act *in vitro* as a good donor for the synthesis of this polymer (Villemez *et al.*, 1965, 1966). It should be noted, however, that other nucleotides could also be of significance as some incorporation occurs with TDP-galacturonic acid in an enzyme system from tomatoes (Lin *et al.*, 1966). This possibility is even more likely as the latter nucleotide was isolated from sugar beet and, in addition, this plant also contains enzymes capable of converting TDP-glucose to TDP-galacturonic acid (Katan and Avigad, 1966a,b).

Another feature of the pectic substances is that they contain various neutral sugars, besides polygalacturonic acid, in part as side chains or as blocks within the main chain. Almost no experimental evidence with regard to the biosynthesis *in vitro* of these pectic fractions is available, which may be due to the difficulties encountered in the analysis of the products in trace amounts.

With UDP-arabinose-^{14}C as a donor, for example, a mixture of heterogeneous polysaccharides is formed by a particulate enzyme preparation from mung bean shoots (Ozduck and Kauss, 1972). Minor fractions from this mixture show some properties of pectic substances but are hard to distinguish from hemicelluloses as the acidic parts of the molecules are not labelled.

B. INTRODUCTION OF THE METHYL ESTER GROUPS

The only detailed biosynthetic study *in vitro* on a component of pectin has been concerned with the origin of the methyl groups which esterify the carboxyl groups of the polygalacturonic acid chains to varying degrees.

Feeding experiments had shown that various ^{14}C-labelled compounds can contribute *in vivo* to the pectin methyl ester groups (Table I). The structural heterogeneity of these compounds indicates that they presumably feed into a common C-1 pool. This includes methyl galacturonate which is obviously not incorporated *in toto* but only after cleavage of the ester group. The fact that compounds which definitely have to be reduced to the level of methanol are incorporated into pectin suggests that N^5-methyltetrahydrofolic acid may be a cofactor in this part of the metabolism and the subsequent pathway most probably leads, via methionine, to S-adenosyl L-methionine (SAM). This latter compound is in turn the immediate methyl donor *in vitro* for esterification of pectin carboxyl groups as well as for the formation of the 4-O-methyl-ether groups in hemicelluloses (see p. 198).

TABLE I

Sources of methyl groups and their incorporation into plant cell wall polysaccharides

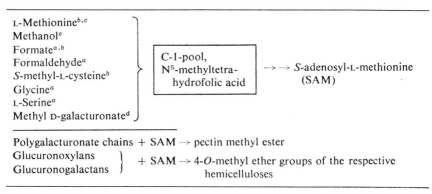

L-Methionine[b,c]
Methanol[e]
Formate[a,b]
Formaldehyde[a]
S-methyl-L-cysteine[b]
Glycine[a]
L-Serine[a]
Methyl D-galacturonate[d]

C-1-pool,
N^5-methyltetra-
hydrofolic acid

→ → S-adenosyl-L-methionine
(SAM)

Polygalacturonate chains + SAM → pectin methyl ester
Glucuronoxylans
Glucuronogalactans } + SAM → 4-O-methyl ether groups of the respective
hemicelluloses

[a] Wu and Byerrum (1958).
[b] Mae et al. (1972).
[c] Sato et al. (1958).
[d] Loewus (1971).
[e] Roberts et al. (1967).

Villemez et al. (1966) showed that labelled UDP-galacturonic acid methyl
ester could not act as a donor for methylated uronic acid groups with a particulate enzyme preparation from mung bean which was able to polymerize
the corresponding unesterified nucleotide to the polygalacturonic acid
backbone of pectin. Essentially the same particulate enzyme preparation
was used to study the biosynthesis of the methyl ester groups from SAM
(Kauss et al., 1967; Kauss and Hassid, 1967b). Some properties of the enzyme
responsible for this transmethylation are summarized in Table II.

One additional important feature is that no acceptor molecules had to be
added to the enzyme preparation. Addition of polygalacturonic acid or pectin

TABLE II

Some properties of the transferase responsible for the formation of polygalacturonate methyl ester from S-adenosyl L-
methionine (SAM)[a]

Optimal pH value	6·8
+ Mg^{2+}, Co^{2+}, Mn^{2+}	no effect
+ Zn^{2+}, Ni^{2+}	inhibition
+ EDTA	30% stimulation
N^5-Methyltetrahydrofolic acid	not a donor
Apparent K_m for substrate (SAM)	6×10^{-5} M

[a] From Kauss and Hassid (1967b).

did not influence the transmethylation reaction, even if they were prepared from the same tissue. Bailey *et al.* (1967) had previously shown that the particulate preparation contained some polygalacturonate which obviously was in some special association with the methyl transferase. The amount of polygalacturonate present, however, appeared to be insufficient to saturate the methyl transferase. This was shown by experiments in which polygalacturonate was generated within the particles by preincubation with UDP-galacturonic acid (Fig. 1). All remaining nucleotide was then destroyed with phosphodiesterase (PDE) and the methylation was started with labelled SAM. The rate of pectin methyl ester formation was higher in sample B which initially contained UDP-galacturonic, as compared with control A. The polymer preformed *in situ* appeared to act as an additional acceptor of methyl groups whereas polymer added exogenously had no access to the methyl transferase.

Figure 1 illustrates a further point. In the case of the control A, the particulate enzyme was treated with enough PDE to destroy any trace of

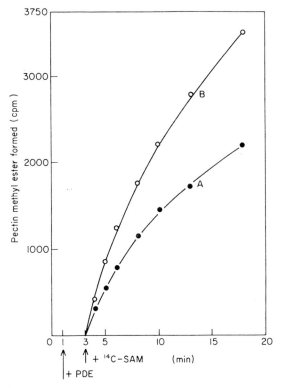

FIG. 1. Time course of the incorporation of methyl ester groups into pectin by particulate enzyme preparations from mung bean shoots preincubated with (B) or without (A) UDP-galacturonic acid followed by phosphodiesterase (PDE). SAM = S-adenosyl L-methionine. (From Kauss and Swanson, 1969.)

nucleotide sugar still present from the original cell. In other experiments the depletion of any endogenous nucleotides which could have served as a primary acceptor of the methyl groups was guaranteed by preincubation of the particles at reaction temperature for prolonged periods (Kauss and Swanson, 1969). This together with the findings in the case of polygalacturonate acceptor molecules formed *in situ*, suggests that methylation in pectin biosynthesis occurs not at the nucleotide level but most likely at the macromolecular level. Although polymerization and methylation are separate reactions, they obviously occur closely together in a structural "compartment" which contains the necessary enzymes as well as the products, i.e. pectic molecules at different stages of formation. Some properties of this compartment can be deduced from the observation that addition of soluble pectin methyl esterase did not have any influence on the amount of ^{14}C-methylated pectin formed during the reaction (Kauss *et al.*, 1969). If, however, the methylated pectin was isolated from the inactivated particles and then added to a similar incubation mixture it was completely hydrolysed in a short time. We could show, in addition, that the particulate enzyme preparation itself contained a potent pectin methyl esterase. Obviously the esterases cannot act on the pectic substances formed in the particles because they remain in a protected form in the "compartment".

The nature of the protecting factor was demonstrated by experiments in which it was overcome by the use of agents capable of destroying lipid membranes (Fig. 2). For this purpose the ^{14}C-methyl-labelled pectin was

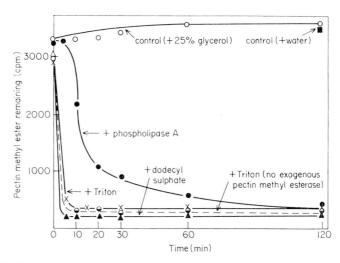

FIG. 2. Influence of detergents and phospholipase A on the hydrolysis, by exogenous pectin methyl esterase, of ester groups formed within a particulate enzyme preparation from mung bean shoots. ^{14}C-Methyl-pectin was generated in the particles by preincubation with S-adenosyl L-methionine-^{14}C and UDP-galacturonic acid (10 min) followed by centrifugation. (From Kauss *et al.*, 1969.)

generated from *S*-adenosyl L-methionine in the particles which were then collected by centrifugation and resuspended with or without soluble pectin methyl esterase. The labelled material in both cases can be incubated for 2 h at 30° without any loss of methyl groups. These groups were readily liberated however, if the integrity of the particles was disturbed by the detergents Triton-X-100 or dodecyl-sulphate or the enzyme phospholipase A. It can be concluded, therefore, that the pectic substances after biosynthesis remain in a structural compartment formed from lipid membranes which in addition contains all the necessary enzymes for polymerization and methylation. These biochemical findings correlate well with those obtained by cytological methods (Northcote, this volume, Chapter 11). Thus the particulate enzyme preparation used for the studies on pectin biosynthesis appears to be derived from the Golgi apparatus or its maturing vesicles. After destruction of the membrane-coated particles by treatment with detergents or phospholipase A, pectin labelled with ^{14}C-galacturonic acid can still be centrifuged out, possibly indicating that within the particles it is still somehow bound to the structure, possibly in a protopectin-like form (Kauss *et al.*, 1969).

IV. BIOSYNTHESIS OF HEMICELLULOSES

A. FORMATION OF THE SUGAR POLYMERS

The term *hemicellulose* is a very trivial name for a vast mixture of hetero-polysaccharides mainly composed of xylose, arabinose, galactose, mannose, glucose and glucuronic acid. If one goes through the literature one can find reports that all of these sugars can be incorporated from their respective isotopically labelled nucleoside diphosphate donors into polymers which are classified as hemicelluloses mainly on the basis of solubility properties (for references see Lamport, 1970; Kauss, 1970). Two major problems arise from such studies. First, one obtains very small quantities of material in which only the labelled monomer can be analysed in detail. This means that it is only possible to gain precise information concerning the types of linkages formed between the newly added sugar residues and the acceptor sites of the polymer. In some cases, when oligosaccharides result from partial acid hydrolysis, the nature of the next few sugar monomers can also be ascertained but the full structure of the macromolecule remains unknown, however. In addition, the situation is complicated by the fact that, in most cases, a full spectrum of chemically similar polysaccharides is synthesized which may have solubilities ranging from water soluble to alkali insoluble; such was the case when GDP-mannose was used as a donor. This problem seems to result from the unfortunate but important fact that in practically all of the relevant studies the ultimate acceptors for the new residues were already present in the enzyme preparations and most likely in the form of macromolecules rather than as low molecular weight primers. It is uncertain, therefore, whether in these cases chains were really growing to a significant extent or whether only a few additional monosaccharide residues were being added simultaneously at many sites.

The second major problem is that the particulate preparations used in incorporation studies contain enzymes for the interconversion of the nucleotide sugars as well as those catalysing polymer formation. Arabinose-containing polysaccharides, for example, can be isolated after incubation of these preparations with labelled UDP-xylose, UDP-glucuronic acid or UDP-galacturonic acid as well as with UDP-arabinose (see Odzuck and Kauss, 1972). This difficulty can be overcome to some extent by using short-time incubations, but then only very limited amounts of material for analytical purposes can be obtained. As it appears to be impossible to remove the interfering enzymes responsible for nucleotide interconversion we have recently tried another approach. Working with a mung bean enzyme preparation we were able to study, separately, the transferases responsible for the incorporation of arabinose and xylose residues (Odzuck and Kauss, 1972). After repeated washing of the particles with Triton-X-100 only the arabinose transferase remained intact. Thus the properties of this enzyme could be studied without interference from the xylose transferase. One of the striking properties of the arabinose transferase is its strict dependence on the presence of Mn^{2+} in optimal concentrations. The xylose transferase, on the other hand, does not need divalent ions; it can, therefore, be examined in the presence of EDTA which under certain conditions fully inactivates the arabinose transferase so that an almost exclusive transfer of ^{14}C-xylose from UDP-xylose-^{14}C to macromolecular material occurs.

B. INTRODUCTION OF THE 4-*O*-METHYL ETHER GROUPS INTO THE GLUCURONIC ACID SIDE CHAINS

The difficulties mentioned above in investigating the incorporation of monosaccharides into hemicellulosic polysaccharides prompted us to work on quite a different aspect of hemicelluloses biochemistry, namely, the origin of the methyl ether group attached to C-4 of the glucuronic acid residue. In independent studies with enzyme preparations from developing corn cobs we had shown that UDP-glucuronic acid could serve as a source of glucuronic acid units in hemicellulose B (Kauss, 1967). During these studies we became aware of the fact that in growing tissues such as corn cobs, grass coleoptiles or mung bean hypocotyls, only a small number of the glucuronic acid residues in the hemicelluloses possessed 4-*O*-methyl ether groups.

We were later able to show that the methyl ether groups were produced when particulate enzyme preparations from these tissues were incubated with methyl-labelled SAM (Kauss and Hassid, 1967a; Kauss, 1969b). Some properties of the hemicellulose methyl transferase involved in this reaction

TABLE III

Some properties of the transferase responsible for the incorporation of methyl ether groups from S-adenosyl L-methionine into hemicellulose B[a]

Optimal pH value	8·0
+ Mg^{2+}	no effect
+ Co^{2+}	4-fold stimulation
+ Mn^{2+}	2-fold stimulation
+ EDTA	inhibited; reversed by Co^{2+} or Mn^{2+}
+ Ni^{2+}, Zn^{2+}	no effect

[a] From Kauss and Hassid (1967a), Kauss (1969b) and Kauss, unpublished results.

are summarized in Table III. The enzyme is clearly different from the pectin methyl transferase (see p. 193). It should also be pointed out that with the former enzyme, methylation is not dependent on the simultaneous presence of SAM and nucleoside diphosphate glucuronic acid. This fact was established as in the case of biosynthetic studies on pectic substances, by treatment of the enzyme system with PDE or by prolonged incubation to allow endogenous nucleotide-utilizing enzymes to act. It appears very likely, therefore, that the methyl group acceptors were again preformed macromolecules present in the enzyme preparation. A less likely alternative explanation would be that the methyl groups were first transferred from S-adenosyl L-methionine to a short chain polysaccharide intermediate, possibly lipid-bound. Examples of the formation of sugar polymers by such a mechanism are known in the case of the biosynthesis of the O-antigen of *Salmonella* (Robbins *et al.*, 1967) and possibly of a glycoprotein in mammalian liver (Parodi *et al.*, 1972). To date, however, no experimental evidence has been published which would unequivocally support the proposition that a similar mechanism exists for the biosynthesis of plant cell wall polysaccharides.

The best way to get information on the structure of the product formed in the methylation reaction is to perform partial acid hydrolysis. The composition of the resulting labelled uronic acid mixture is different when different enzyme sources are used (Kauss, 1969b). From both mung bean and corn cob enzyme preparations some free 4-O-methylglucuronic acid is liberated, but the predominant aldobiouronic acid from the mung bean system is 4-O-methyl-glucuronogalactose and from the corn cob 4-O-methylglucuronoxylose. This means that the polysaccharides methylated by the transferase are mainly glucuronogalactans in mung bean and glucuronoxylans in corn cob preparations which appears to reflect the proportion of the main hemicellulose B components present in cell walls of the Leguminosae as compared with maize. The results obtained from the studies of the origin of 4-O-methyl-

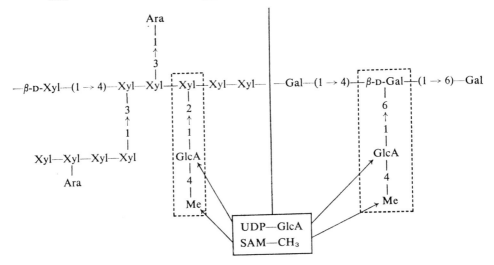

FIG. 3. Biosynthesis of the 4-O-methyl-glucuronic acid side chains of hemicellulose B cell wall components based on xylans (maize) and galactans (mung bean). The dotted lines enclose the major aldobiouronic acids released from the two types of polymer by partial acid hydrolysis. SAM = S-adenosyl L-methionine.

glucuronic acid side chains of hemicelluloses are summarized in Fig. 3. Here idealized structures of sectors of the two types of hemicellulose B, based on xylan or galactan, are given. The experiments have clearly shown that the methyl ether groups are not incorporated into polysaccharides via a methylated glucuronic acid nucleotide derivative. The donor for glucuronic acid residues in the hemicelluloses is UDP-glucuronic acid. Of general importance is the fact that in the case where glucuronic acid was transferred to a galactan, no substrate (e.g. UDP-galactose) was present which would have allowed a simultaneous polymerization of the basic galactan chain. This may indicate that the new glucuronic acid residues are also incorporated into preformed polysaccharide acceptors as appears to occur with the methyl ether groups originating from S-adenosyl L-methionine.

V. LIPID INTERMEDIATES

In past years convincing evidence has accumulated that in the biosynthesis of some polymers special glycolipids are involved as intermediates. Some important features of the well established systems from bacteria, yeast and mammals are summarized in Table IV. Section A shows two different types of phosphorylated glycolipids which are known to occur and their modes of synthesis. In the first case, a lipid monophosphate reacts with a nucleoside diphosphate sugar and a sugar 1-phosphate residue is transferred, leading to the formation of the corresponding nucleoside monophosphate and a lipid-pyrophosphate-sugar derivative. The latter type of compound is known

TABLE IV

Some properties of polyprenol glycolipids known to function as intermediates in the biosynthesis of carbohydrate-containing polymers in bacteria, yeast and mammals

A	L—P	+ N—P—P$\sim$S $\rightleftharpoons$ N—P	+ L—P—P$\sim$S^a
	L—P	+ N—P—P$\sim$S $\rightleftharpoons$ N—P—P +	L—P$\sim$S
	Phospholipid	*Nucleoside diphosphate sugar*	*Phosphorylated glycolipid*

B $\quad$ L—P—P$\sim$S′ + N—P—P$\sim$S″ → L—P—P$\sim$S′—S″ + N—P—P

$$\text{CH}_3$$
$$|$$

C $\quad$ H—[CH$_2$—C=CH—CH$_2$]$_x$—OH

bacteria:	$x = 11$	(undecaprenol; Lennarz and Scher, 1972)
yeast:	$x = 14-18$	(dolichols; Jung and Tanner, 1973)
mammalian liver:	$x =$ mainly 20	(dolichols; Parodi *et al.* 1972)
higher plants:	$x = 6-13$	(betulaprenols, ficaprenols; Wellburn and Hemming, 1966)

^{a} L = lipid; N = nucleoside; S = glycosyl; P = phosphate.
Variations in the number of double bonds and the steric relationships of the methyl groups are not shown.

to be involved in the biosynthesis of the *Salmonella O*-antigen (Robbins *et al.*, 1967) and of the bacterial cell wall peptidoglycan (Strominger *et al.*, 1972). The second type of phosphorylated glycolipid has been well studied, for example, in *Micrococcus* (Lennarz and Scher, 1972; this reference also gives an excellent review of most of the other polyprenol sugar intermediates) and in liver (Parodi *et al.*, 1972). It is formed by transfer of sugar only, from nucleoside diphosphate sugar to phospholipid and the products are, therefore, a nucleotide diphosphate and a lipid-monophosphate-sugar. Both types of phosphorylated glycolipid exhibit sugar transfer potentials of the same order as the nucleotide diphosphate sugars (as indicated by the reversibility of the reactions in Section A) and can, therefore, function as sugar donors in the hydrophobic environment of membranes. In addition (Section B, Table IV) oligosaccharides, or in some cases short chain polysaccharides, can be built up on the lipid by transfer of glycosyl residues from nucleoside diphosphate sugars. These sugar chains can then be transferred to final acceptors which are mostly complex polymers. In the two systems known in eukaryotic cells, i.e. the yeast (Tanner *et al.*, 1972) and the liver (Parodi *et al.*, 1972) systems, the final products are glycoproteins.

The lipid moieties of all of these glycolipid intermediates are polyprenol alcohols composed of varying numbers of isoprenoid units (see Table IVC). Polyprenols, up to the present time, have only been isolated in the free form from higher plants. Their physiological function may be to form intermediates for the biosynthesis of plant polysaccharides and this will be discussed in the last part of this chapter.

In 1964 Colvin predicted that lipophilic carriers played a role in cellulose biosynthesis although there was no satisfactory biochemical evidence to support this idea. During the time when the bacterial lipid intermediates were being explored we made a survey of the plant particulate enzyme preparations which synthesized polysaccharides from various sugar nucleotides and looked for the simultaneous formation of glycolipids. Although we obtained small amounts of labelled lipids from some nucleotides their properties did not suggest that they were polysaccharide intermediates. The lipids formed from UDP-arabinose and UDP-glucose, for example, were not hydrolysed by 0·01 N hydrochloric acid and, therefore, it appeared unlikely that they would exhibit a high sugar transfer potential. This was especially true for the lipid material produced from UDP-glucose. This lipid was composed of 90–98% free and acylated steryl glucosides and did not show turnover under "chase conditions" (Kauss, 1968). The identity and function of the few remaining percent was difficult to evaluate. The only lipid where we had enough material for detailed studies and which turned out to have the properties of an intermediate was that formed from GDP-mannose in mung bean preparations under conditions of mannan formation (Kauss, 1969a). The formation of mannan by this preparation proceeds by a reaction proportional to time for periods up to 5 min, whereas, in contrast, the pool of ^{14}C-mannosyl-lipid seems to be rather small as it can be saturated within seconds with label. The turnover of ^{14}C-mannosyl units in the lipid can be clearly demonstrated by "chase" experiments with unlabelled GDP-mannose.

The mannosyl-lipid was extracted from the particulate fraction with water-saturated butanol and purified by various techniques. It showed chromatographic properties similar to those of the polyprenol-monosphate-mannose isolated by Scher *et al.* (1968) from *Micrococcus*. The high sugar transfer potential of the glycolipid is indicated by two properties. All of the mannose was liberated from the lipid by hydrolysis with 0·01 N hydrochloric acid at 100°C within 20 min, which is comparable to the rate of hydrolysis of GDP-mannose. Additional direct evidence can be drawn from experiments where purified ^{14}C-mannose-lipid was incubated with the particulate enzyme preparation. When GDP was present as an acceptor a considerable amount of GDP-^{14}C-mannose was synthesized by reversal of the reaction by which the mannosyl-lipid had been originally formed. The reversibility of reaction shows that the group transfer potential of both the mannosyl-lipid and GDP-mannose must be of the same order of magnitude and allows the conclusions that the mannosyl-lipid may function as a donor of mannose units and has the structure of a mannosyl-monophosphate-lipid. The occurrence and the general properties of the mannosyl-lipid from mung beans has been confirmed by Villemez (1970) as well as by Storm and Hassid (1972). A first indication of the isoprenoid nature of the lipid part of the molecule was given by the finding that it could be derived from ^{3}H-mevalonic acid *in vivo* (Kauss, 1969a). It was shown later (Alam and Hemming, 1971; Forsee and Elbein, 1973) that the lipid part of the mannosyl-lipid is most likely a straight-chain poly-prenol of the ficaprenol-type. It is of interest to note that cotton fibres contain a similar glucosyl-monophosphate-polyprenol (Forsee and Elbein, 1973).

Unfortunately, the physiological significance of the intermediate is not fully apparent. It was difficult to demonstrate the transfer of mannose from the glycolipid to polymer probably because with the particles suspended in aqueous solutions insufficient glycolipid reached the polymerizing enzymes in the membranes. Under certain conditions, and not with every experiment, we finally observed a small but definite time-dependent incorporation of labelled mannose from the lipid into a polymer. One important observation was that incorporation was only possible when unlabelled GDP-mannose was present in addition to the labelled lipid (Kauss, 1972). The mannose linkage in the polymer was different from the linkage in the lipid as 1 N acid for 3 h at 100°C was needed to hydrolyse it. The product could not be characterized in more detail as only small amounts were available.

At the moment we feel that if the mannosyl-lipid has any role *in vivo* then it is only concerned with special functions, possibly the synthesis of side chains of a mannan-like material or a glycoprotein; its involvement as an intermediate in the synthesis of the bulk of the mannans or glucomannans is unlikely.

To date, as far as I see, there is no published experimental evidence which would allow one to suppose that the concept of the lipid intermediate can be applied to formation of the numerous other hemicellulose polysaccharides. A report on the formation of "intermediates" by enzyme preparations capable of polysaccharide biosynthesis with UDP-glucose as a donor (Ville-mez *et al.*, 1968) was prompted by the demonstration that the lipophilic material was a mixture of free and acylated sterol glucosides which did not show transfer properties in chase experiments and were not acid labile (Kauss, 1968). The role of another phenol-soluble material formed from GDP-mannose-^{14}C by mung bean preparations (Villemez and Clark, 1969) as an intermediate in the formation of polysaccharides is doubtful. These latter experiments still await confirmation by other investigators and more pub-lished data by the original authors before the transfer properties of this material can be fully assessed.

REFERENCES

Alam, S. S. and Hemming, F. W. (1971). *FEBS Letters* **19**, 60.
Bailey, R. W., Haq, S. and Hassid, W. Z. (1967). *Phytochemistry* **6**, 293.
Bolker, H. J. and Wang, P. Y. (1969). *Tappi* **52**, 920.

Cleland, R. (1971). *A. Rev. Pl. Physiol.* **22**, 197.
Colvin, J. R. (1964). *Can. J. Biochem. Physiol.* **39**, 1921.
Jung, P. and Tanner, W. (1973). *Eur. J. Biochem.* **37**, 1.
Jung, P., Tanner, W. and Wolter, K. (1972). *Phytochemistry* **11**, 1655.
Forsee, W. T. and Elbein, A. D. (1973). *J. biol. Chem.* **248**, 2858.
Katan, R. and Avigad, G. (1966a). *Israel J. Chem.* **3**, 110.
Katan, R. and Avigad, G. (1966b). *Biochem. biophys. Res. Commun.* **24**, 18.
Kauss, H. (1967). *Biochim. biophys. Acta* **148**, 572.
Kauss, H. (1968). *Z. Naturf.* **23b**, 1522.
Kauss, H. (1969a). *FEBS Letters* **5**, 81.
Kauss, H. (1969b). *Phytochemistry* **8**, 985.
Kauss, H. (1970). *In* "Fortschritte der Botanik" (H. Ellenberg *et al.*, eds), Vol. 32, p. 69. Springer-Verlag, Berlin.
Kauss, H. (1972). *In* "Biochemistry of the glycosidic linkage" (R. Piras and H. G. Pontis, eds), PAABS Symposium, Vol. 2, p. 221. Academic Press, New York and London.
Kauss, H. and Hassid, W. Z. (1967a). *J. biol. Chem.* **242**, 1680.
Kauss, H. and Hassid, W. Z. (1967b). *J. biol. Chem.* **242**, 3449.
Kauss, H. and Swanson, A. L. (1969). *Z. Naturf.* **24b**, 28.
Kauss, H., Swanson, A. L. and Hassid, W. Z. (1967). *Biochem. biophys. Res. Comm.* **26**, 234.
Kauss, H., Swanson, A. L., Arnold, R. and Odzuck, W. (1969). *Biochim. biophys. Acta* **192**, 55.
Lamport, D. T. A. (1970). *A. Rev. Pl. Physiol.* **21**, 235.
Lennarz, W. J. and Scher, M. G. (1972). *Biochem. biophys. Acta* **265**, 417.
Lin, T. Y., Elbein, A. D. and Su, J. C. (1966). *Biochem. biophys. Res. Comm.* **22**, 650.
Loewus, F. (1965). *Fed. Proc.* **24**, 855.
Loewus, F. (1971). *A. Rev. Pl. Physiol.* **22**, 337.
Mae, T., Ohira, K. and Fujiwara, A. (1972). *Pl. Cell Physiol.* **13**, 407.
Nikaido, H. and Hassid, W. Z. (1971). *Adv. Carbohyd. Chem. Biochem.* **26**, 351.
Northcote, D. H. (1972). *A. Rev. Pl. Physiol.* **23**, 113.
Odzuck, W. and Kauss, H. (1972). *Phytochemistry* **11**, 2489.
Parodi, A. J., Behrens, N. H., Leloir, L. F. and Carminatti, H. (1972). *Proc. natn. Acad. Sci. U.S.A.* **69**, 3268.
Robbins, P. W., Bray, D., Dankert, M. and Wright, A. (1967). *Science, N.Y.* **158**, 1536.
Roberts, R. M., Shah, R. H., Golebiewski, A. and Loewus, F. (1967). *Pl. Physiol.* **32**, 1732.
Sato, C. S., Byerrum, R. U., Albersheim, P. and Bonner, J. (1958). *J. biol. Chem.* **233**, 128.
Scher, M., Lennarz, W. J. and Sweeley, C. C. (1968). *Proc. natn. Acad. Sci. U.S.A.* **59**, 1313.
Storm, D. L. and Hassid, W. Z. (1972). *Pl. Physiol.* **50**, 473.
Strominger, J. L., Higashi, T., Sandermann, H., Stone, K. J. and Willoughby, E. (1972). *In* "Biochemistry of the glycosidic linkage" (R. Piras and H. G. Pontis, eds), PAABS Symposium, Vol. 2, p. 135. Academic Press, New York and London.
Tanner, W., Jung, P. and Linden, J. C. (1972). *In* "Biochemistry of the glycosidic linkage" (R. Piras and H. G. Pontis, eds), PAABS Symposium, Vol. 2, p. 227. Academic Press, New York and London.

Villemez, C. L. (1970). *Biochem. Biophys. Res. Commun.* **40**, 636.
Villemez, C. L. and Clark, A. F. (1969). *Biochem. biophys. Res. Commun.* **36**, 57.
Villemez, C. L., Lin, T. Y. and Hassid, W. Z. (1965). *Proc. natn. Acad. Sci. U.S.A.* **54**, 1626.
Villemez, C. L., Swanson, A. L. and Hassid, W. Z. (1966). *Archs Biochem. Biophys.* **116**, 446.
Villemez, C. L., Vordak, B. and Albersheim, P. (1968). *Phytochemistry* **7**, 1561.
Wellburn, A. R. and Hemming, F. W. (1966). *Phytochemistry* **5**, 969.
Wu, P. H. L. and Byerrum, R. U. (1958). *Pl. Physiol.* **33**, 230.

CHAPTER 14

Biosynthesis of Algal Polysaccharides

ARNE HAUG and BJØRN LARSEN

Institute of Marine Biochemistry, University of Trondheim,
Trondheim, Norway

I. INTRODUCTION

Algae are essentially photosynthetic organisms containing chlorophyll *a* which are distinguished from higher plants mainly by the specialization of the latter for life in a terrestrial environment. In the sea, algae are the dominating primary producers, and they also play an important role in other aquatic environments.

The algae are a very heterogeneous group of plants; they are classified into a number of divisions which often have only remote relationships to each other. The chemist or biochemist studying these organisms should, therefore, not expect to find similarities between two organisms just because both are algae; significant biochemical similarities between the different algal divisions, on the other hand, might be of considerable interest from a phylogenetic point of view.

As might be expected, a wide variety of polysaccharides have been isolated from algae. It is often convenient when discussing plant polysaccharides to divide them into three main groups: (1) storage polysaccharides; (2) the fibrillar components of the cell walls and (3) the matrix components of the cell walls (including extracellular polysaccharides the function of which is still obscure). The storage polysaccharides of algae are mainly glucans and these fall into two main classes: (a) those with a backbone of α-$(1 \rightarrow 4)$-linked D-glucose residues with side-chains joined by α-$(1 \rightarrow 6)$ linkages to the main chain and (b) those with β-(1 → 3)-linked glucose residues in the main chain and β-(1 → 6)-linked side-chains. The fibrillar cell wall components are in most cases cellulose, but some well established exceptions are found among Chlorophyta (green algae) and Rhodophyta (red algae), where, in some cases, these fibrillar components are either β-(1 → 4)-linked mannan or β-(1 → 3)-linked xylan. The third group, the matrix polysaccharides of the cell walls, comprises a wide variety of glycans about which few generalizations can be made. However, they normally possess either carboxyl groups, sulphate half-ester groups or both, which often results in the polymers having a high negative charge density; other substituents also commonly occur and in particular, methyl groups.

The biosynthesis of polysaccharides involves first the formation of the appropriate precursors, secondly, the polymerization process and, finally in some cases, modifications of the polysaccharide molecule by substitution or other reactions. In the following discussion these three steps will be considered separately.

II. Precursors in Polysaccharide Biosynthesis

The biosynthesis of polysaccharides in marine algae involves the incorporation of a number of neutral and acidic monosaccharides into the polymer. These include the ubiquitous D-glucose and also D-galactose, D-mannose, L-galactose, L-rhamnose, L-fucose, D-xylose, D-mannuronic acid, D-glucuronic acid and L-guluronic acid as well as some other sugars. Few attempts have been made to isolate and characterize the actual precursors taking part in the biosynthetic polymerization processes. With our present state of knowledge in the field of polysaccharide biosynthesis, it is reasonable to expect these precursors to be glycopyranosyl esters of nucleoside pyrophosphates (Nikaido and Hassid, 1971). The term sugar nucleotides will be used throughout the following discussion to describe these molecules, and the usual capital letters for the different nucleotide bases will be applied.

Three different glucosyl nucleotides have been reported in green algae: ADP-D-glucose in *Chlorella* (Kauss and Kandler, 1962) and UDP-D-glucose and GDP-D-glucose in *Asteromonas* (Pakhomova and Zaitzeva, 1967). ADP-D-glucose was produced from ATP and α-D-glucose-1-phosphate using a *Chlorella* ADP-glucose pyrophosphorylase (Sanwal and Preiss, 1967). In the red algae *Porphyra perforata*, Su and Hassid (1962) demonstrated the presence of UDP-D-glucose and, in addition, UDP-D-glucuronic acid, GDP-D-mannose and GDP-L-galactose. The isolation of the latter component is particularly interesting as this presumably is the precursor of the L-galactose units characteristic of the *Rhodophyta* galactans of the agar-porphyran type. In the other main group of *Rhodophyta* galactans, the carrageenans, only the D-enantiomorph of galactose occurs (Percival and McDowell, 1967; Haug, 1974). Su and Hassid suggest that GDP-L-galactose is formed in the algae by an enzyme system capable of carrying out some of the same reactions that occur in the conversion of GDP-D-mannose to GDP-L-fucose. The latter type of enzyme system has been reported to occur in bacteria and higher plants (Nikaido and Hassid, 1971; Liao and Barber, 1971; Glaser and Kornfeld, 1961). In the green alga *Chlorella pyrenoidosa*, Barber (1971) has recently described an enzyme system which is capable of transforming labelled GDP-D-mannose into L-galactose-1-phosphate. It is reasonable to assume that in the intact plant the useful end-product, which would serve as a precursor for polysaccharide biosynthesis, would be GDP-L-galactose. A polysaccharide from *Chlorella pyrenoidosa* containing L-galactose residues has recently been described by White and Barber (1972). The related transformation of UDP-D-glucose into UDP-L-rhamnose was also catalysed by extracts of the same algae (Barber and Chang, 1967). NADPH was an essential requirement for this latter reaction which was shown to proceed via the intermediate, UDP-6-deoxy-D-xylo-4-hexoseulose.

The only study on nucleotides in brown algae has been carried out by Lin and Hassid (1966a). Chromatographic evidence indicated that extracts of *Fucus gardneri* contained UDP-D-glucose, UDP-L-arabinose and a mixture of GDP derivatives of mannose, glucose, arabinose, fucose and possibly galactose. Lin and Hassid were also able to isolate and characterize GDP-D-mannuronic acid together with smaller amounts of another acidic nucleotide which was probably the corresponding L-guluronic acid derivative. The following enzymes were also shown to be present in *Fucus*: hexokinase, phosphomannomutase, D-mannose-1-phosphate guanylyl transferase and GDP-D-mannose dehydrogenase. These could account for the conversion of D-mannose to GDP-D-mannuronic acid *in vivo*. The GDP-L-guluronic acid which is present in the alga could be formed by a C-5-epimerization reaction analogous to the C-5-epimerization of UDP-D-glucuronic acid to UDP-L-iduronic acid which has been shown to occur in rabbit skin (Jacobson and Davidson, 1962). The two GDP-uronic acid derivatives found in *Fucus* could be considered as likely precursors of alginic acid, the main matrix polysaccharide of the brown algae. However, recent work, to be discussed later (see p. 213), has shown that this may not be the case.

Studies with some algae clearly indicate that the organisms contain the sugar nucleotides required as precursors for polysaccharide biosynthesis, and that, like higher plants, they also contain enzyme systems capable of carrying out sugar interconversions at the nucleotide level. Investigations have so far been limited, however, and the sugar nucleotide contents of several families of algae have not been investigated.

In the case of higher plants, Loewus (1965; 1971) has proposed a pathway for uronic acid formation which involves the oxidative cleavage of inositol derivatives. The possibility that such alternative pathways of metabolism, not involving sugar nucleotides, exist in the algae has not been examined.

III. The Polymerization Process

The process of polymerization is essentially a sequence of transfer reactions of monosaccharide units from nucleotide donors to an acceptor, or primer, molecule which is catalysed by specific transferases. A lipid derivative may in some cases serve as a primary acceptor for the monosaccharide units, thus serving as an intermediate carrier for mono- or oligo-saccharide units prior to their incorporation into the polymer molecule (Robbins *et al.*, 1967).

In the following discussion the glucans which function as storage poly-saccharides will first be considered. The synthesis of starch in *Chlorella pyrenoidosa* (Chlorophyta) takes place by the transfer of glucose units from ADP-glucose to an acceptor molecule by means of the enzyme ADP-D-glucose: α-(1 → 4)-glucan: α-4-glucosyl transferase. This enzyme was extracted from disrupted *Chlorella* cells and partially purified by Preiss and Greenberg (1967). The only substrates accepted by the enzyme are ADP-D-glucose ($K_m \sim 2 \cdot 7 \times 10^{-4}$) and dADP-D-glucose ($K_m \sim 4 \cdot 7 \times 10^{-4}$), whereas a number of α-(1 → 4)-glucans can serve as acceptors. The formation of new α-(1 → 4)-linkages was demonstrated by the isolation of [14C]maltose from a hydrolysate of the enzymic product obtained by using ADP-D-[14C]glucose as substrate. The presence of the same type of enzyme which transfers labelled glucose from ADP-D-[14C]glucose to floridean starch has been demonstrated in isolated starch granules from the red alga *Serraticardia maxima* (Nagashima *et al.*, 1971). The enzyme activity could not be separated from the granules, a situation which also occurs with transferases in higher plants. Branching enzyme has been shown to be present in another red alga, *Rhodymenia pertusa* by Fredrick (1971). The available evidence thus indicates that the formation of starch both in green and red algae follows the same general pattern as in higher plants.

Glucans with β-(1 → 3)-linkages occur in several algal classes such as the Phaeophyta (brown algae), Euglenophyta, Chrysophyta and Bacillariophyta (diatoms). They have been given different names (laminaran, paramylon, chrysolaminaran and leucosin, respectively), but they all probably function as storage polysaccharides. The β-(1 → 3)-linked glucan of higher plants, callose, is structurally very similar to algal glucans but apparently has a different function, playing a significant part in the development of sieve plates and being produced after injury to the plant tissue (Shafizadeh and McGinnis, 1971). The biosynthesis of a β-(1 → 3)-glucan was reported by Goldenberg and Marechal (1963) who used an enzyme preparation from *Euglena gracilis* with UDP-D-[14C]glucose as substrate. The enzyme appeared to be fairly specific for UDP-α-D-glucose; the corresponding ADP derivative resulted in a 75% decrease in activity and no activity was observed with UDP-β-D-glucose. The identity of the product was checked by chromato-graphic analysis of the 14C-labelled oligosaccharides resulting from acid hydrolysis of the polymer. The polysaccharide was also degraded by a laminaranase isolated from the same plant. The *Euglena* transferase system was also studied by an Australian group (Droyer *et al.*, 1970) with reference to the synthesis of polysaccharide by the alga under various conditions of growth. A sharp decrease in transferase activity which coincided with the cessation of net synthesis of β-(1 → 3)-glucan was observed.

As mentioned in the Introduction, the structural polysaccharides in algae may conveniently be divided into fibrillar and matrix components. Some green algae, notably *Valonia*, have very well-developed cellulose fibrils and have been used to study cellulose structure and synthesis. Of particular interest in this connection is the work of Marx-Figini (1969a), who showed that most of the cellulose in *Valonia* was monodisperse with a degree of polymerization of 19 000. Similar observations on the cellulose of higher plants have been published by the same author (Marx-Figini, 1969b); in this case, however, the reported degree of polymerization (14 000) was somewhat lower. These observations may indicate that the polymerization process involved in the formation of fibrillar components may, at least in some cases, require a template-like termination mechanism. Marx-Figini (1971) found, however, that the presence of colchicine did not influence the degree of polymerization of the *Valonia* cellulose, and interpreted this as an indication that microtubuli were not involved in the polymerization process.

It is generally accepted, however, that the biosynthesis of polysaccharides in most cases is a non-template process. It is reasonable to assume that this is also the case for the matrix polysaccharides of algae, although there is very little direct evidence available to support this assumption at present. Only in the case of alginate are the biosynthetic routes involved in the formation of the polymer reasonably well characterized. The work by Lin and Hassid (1966a) provided evidence for a pathway from mannose to GDP-D-mannuronic acid, a putative substrate in the polymerization process. Lin and Hassid (1966b) also succeeded in demonstrating the incorporation of ^{14}C into a polyuronide fraction in a cell-free system prepared from *Fucus gardneri*, using GDP-D-[^{14}C]mannuronic acid. Evidence was thus provided for the existence of a mannuronic acid transferase in brown algae. The main remaining problem was the origin of the L-guluronic acid units in alginate. The presence in *Fucus* of small amounts of a nucleotide which was tentatively identified as GDP-L-guluronic acid (Lin and Hassid, 1966a) makes it reasonable to expect that guluronate is also added to the polymer by a transferase-catalysed reaction. An alternative mechanism, involving epimerization of D-mannuronic to L-guluronic acid residues at the polymer level has been suggested, however, and this will be discussed in the following section.

IV. MODIFICATION OF POLYMERS

A high degree of substitution is a characteristic feature of many algal polysaccharides. The most conspicuous of these substituents are perhaps the sulphate half-ester groups, which may change an otherwise neutral polymer into a charged polyanion. This is typically the case in the sulphated galactans of most red algae and the fucose-containing polysaccharides of the brown algae, although in some cases the unsubstituted polymer may also carry charged groups in the form of uronic acid residues. In some of the Rhodophyta galactans (agars) another anionic substituent is present in the form of pyruvic acid moieties linked as ketals to position 4 and 6 of galactose units. A third type of substituent which commonly occurs in the galactans of red algae is the O-methyl group which is most frequently found on the C-6 positions of the galactose units.

We have chosen to consider the substitution reaction under the heading, "Modification of Polymers", thereby indicating that the substitutions occur after the polymerization has taken place. This assumption is to a large extent based on indirect evidence. No substituted precursors of polysaccharides have been reported to occur in the algae and for related polysaccharides in other organisms it is well established in many cases that substitution occurs at the polymer level. In the case of pectin in higher plants, the available evidence indicates that the esterification of the D-galacturonic acid occurs by a transfer of the methyl group from S-adenosyl-L-methionine at a late stage in the synthesis of the polysaccharide (Kauss and Hassid, 1967a). The same type of reaction produces 4-O-methyl-D-glucuronic acid units in hemicelluloses and in this case it was demonstrated that the acceptor molecule was a macromolecular entity (Kauss and Hassid, 1967b). Su and Hassid (1962) suggested that the reaction responsible for the 6-O-methylation of galactose units in the galactan of *Porphyra perforata* occurs by the same mechanism; i.e. at the polymer level. No direct experimental evidence for this was presented, but if it is accepted that UDP-D-galactose and UDP-L-galactose are the precursors in the polymerization reaction, it follows that substitution must occur at a later stage.

Many of the same arguments may be used with regard to sulphate substitution. It is now generally assumed that the sulphate groups of animal proteoglycans are introduced at the polymer level (Stoolmiller and Dorfman, 1969). Experiments with incorporation of $^{35}SO_4^{2-}$ into sulphate galactans of *Chondrus crispus* (Loewus *et al.*, 1971) and *Porphyridium aerugineum* (Ramus and Groves, 1972) demonstrated that the incorporation was a remarkably rapid process. In the unicellular organism, *Porphyridium*, it appeared to be light dependent and the synthesis, movement and deposition of the capsular sulphated polysaccharide was found to be a Golgi-mediated process. The rapidity with which the sulphate incorporation takes place seems to favour the idea that substitution of a preformed polymer occurs. The sulphate is probably activated by the formation of the intermediate, adenosine 3'-phosphate-5'-phosphosulphate (Su and Hassid, 1962).

It is well known that the Rhodophyta galactans often contain 3,6-anhydrogalactose units, the L-enantiomorph in the agar-porphyran family of polysaccharides, and the D-enantiomorph in the carrageenan family (Percival and McDowell, 1967; Haug, 1974). It is also well known that elimination of sulphate from galactose-6-sulphate and the concomitant formation of 3,6-anhydrogalactose units may be achieved by heating the polysaccharide in alkaline solution (Rees, 1961a, 1963). By treating porphyran (the galactan isolated from *Porphyra umbilicalis*) with an enzyme preparation from the same plant, Rees (1961b) demonstrated the formation of 3,6-anhydro-L-galactose residues and the elimination of sulphate. The reaction was essentially irreversible and indicated the presence in the plant of an L-galactose-6-sulphate alkyl-transferase (cyclizing). A similar enzyme isolated from *Gigartina stellata* which reacts in the same way with the carrageenan family of polysaccharides has recently been described (Lawson and Rees, 1970). It is of interest in this connection to note that the gel-forming ability of Rhodophyta galactans increases as the 4-linked galactose-6-sulphate units are converted to 3,6-anhydro-galactose residues. The enzymes just described are thus modifying the matrix polysaccharides of Rhodophyta in the direction of higher gel strength.

In their work on *Porphyra perforata*, Su and Hassid (1962) proposed that, analogous to the intraresidual alkylation coupled to sulphate elimination described above, the methylation of D-galactose residues might also be coupled to sulphate elimination. According to this hypothesis, the backbone is first synthesized and subsequently the C-6 hydroxyl groups of both L- and D-galactose units are esterified with sulphate groups. In the next step, some of the galactose units undergo desulphation which in the case of the D-enantiomorph is accompanied by 6-O-methylation and with the L-enantiomorph leads to formation of 3,6-anhydro-L-galactose. Substitution with sulphate would thus partly function as an activation process prior to further enzymic modification of the polymer.

Another example of modifications taking place at the polymer level is provided by alginate. Alginate is the major matrix polysaccharide of brown algae but has recently also been found to be an extracellular product of certain bacteria (Linker and Jones, 1966; Gorin and Spencer, 1966). The extracellular polysaccharide produced by *Azotobacter vinelandii* was found to have the typical block structure (see p. 215) of algal alginates (Haug *et al.*, 1968; Larsen and Haug, 1971a) and could only be distinguished from the latter by the presence of acetyl groups. The work of Lin and Hassid (1966b) made it reasonable to assume that the polysaccharide chain was formed by a stepwise transfer of the two monomers from nucleotide precursors. To our surprise, however, we observed that the polymer in the supernatant solution obtained after the removal of bacteria, under certain conditions underwent modification which resulted in a higher guluronic acid content. From the same supernatant solution we were able to isolate a protein fraction by ammonium sulphate precipitation, which was capable of converting poly-mannuronic acid, isolated from the brown alga *Ascophyllum nodosum* (Haug *et al.*, 1968), to a typical block copolymer of D-mannuronic and L-guluronic acid which was indistinguishable from an ordinary algal alginate (Larsen and Haug, 1971a; Haug and Larsen, 1971). Incorporation of tritium into the polymer took place when the reaction was allowed to proceed in tritiated water and more than 90% of the label was located in the guluronic acid units (Larsen and Haug, 1971b). The conclusion drawn from these experiments was that a C-5 epimerization reaction took place at the polymer level which converted "in-chain" D-mannuronic acid units to L-guluronic acid units. The results also showed that the polymannuronic acid 5-epimerase led to the formation both of homopolymeric L-guluronic acid blocks and blocks of an alternating structure and that both these blocks were apparently end-products of the epimerization, i.e. that no conversion of alternating blocks into homopolymeric guluronic acid blocks took place (Haug and Larsen, 1971). No epimerization of L-guluronic acid units was observed.

The enzyme depends upon the presence of calcium ions for its activity but does not require NAD^+ or NADH. Our proposition (Larsen and Haug, 1971b), based mainly upon the tritium incorporation studies and the lack of effect of NAD/NADH, is that the first step in the epimerization is an abstraction of H-5 from the mannuronic acid residue. It seems reasonable to assume that this type of epimerization is confined to molecules where the hydrogen atom to be abstracted is adjacent to a carboxyl or a carbonyl group.

Since the normal substrate for this epimerase is a polymer, the dependence of the activity of the enzyme on the degree of polymerization (DP) of the substrate is of particular interest. The activity was found to be independent of DP down to values of 15 and to be very low or absent when the DP of the substrate was 10 or less (Larsen and Haug, 1972). A very sharp decrease in the efficiency of the enzyme thus occurs within a narrow range in the chain length of the substrate.

These results strongly indicate that the biosynthetic pathway to alginate in bacteria involves the formation of polymannuronic acid followed by C-5 epimerization of some of the mannuronic acid residues at the polymer level, leading to two different types of L-guluronic acid-containing blocks. Whether the biosynthesis of alginates in brown algae follows the same pathway is not yet known. ^{14}C-incorporation studies with $NaH^{14}CO_3$ showed that there was an increase in radioactivity of the guluronic acid units in periods when the plants were kept in darkness and where very little net synthesis of alginate took place (Hellebust and Haug, 1972). These results could be easily explained by assuming an epimerization of D-mannuronic acid to L-guluronic acid on the polymer level but the results are not clear enough to furnish proof of this pathway.

If we assume that epimerization at the polymer level does take place in algae then alginate furnishes a second example of a modification of matrix polysaccharides in these organisms in the direction of higher gel strength. In this case an increase in the L-guluronic content of the alginates leads to an increased affinity for calcium ions (Smidsrød and Haug, 1968) and a a higher gel strength of the calcium alginate (Smidsrød and Haug, 1972).

The reaction described above was the first reported example of epimerization of sugar residues in a preformed polysaccharide molecule. Recently, however, Lindahl et al. (1972) showed that L-iduronic acid units in heparin were formed by epimerization of D-glucuronic acid units in the polymer. In dermatan sulphate, epimerization at the polymer level in the opposite direction, from L-iduronic to D-glucuronic acid, has recently been detected (Fransson et al., 1974). Stereochemical modification of glycosyl units in a polysaccharide chain does, therefore, appear to be of widespread occurrence.

V. Sequence of Monomeric Units in Heteropolysaccharides Synthesized by a Non-template Process

As discussed above, two different biosynthetic routes may be proposed for alginate, one involving the usual transfer of monomers from nucleotide precursors to the growing polymer chain and the other an epimerization reaction at the polymer level which changes a homopolymer into a heteropolymer. Regardless of which of these two mechanisms is responsible for the formation of the finished heteropolysaccharides, the sequence of the two monomeric units along the polymer chain merits further discussion and may serve to illustrate some of the general problems involved in the formation of heteropolysaccharides by non-template processes.

Chemical studies have shown that alginic acid may best be described as a block copolymer consisting of three types of blocks; homopolymeric blocks of D-mannuronic acid units and of L-guluronic acid units and blocks of a predominantly alternating sequence of the two monomers (Haug, 1974; Haug et al., 1966; Haug et al., 1967). The sequence of the two monomers in a heteropolymer synthesized by a non-template process is a result of the specificity of the enzymes taking part in the formation of the polymer. This may most conveniently be discussed by considering a polymerization process where two types of monomers (A and B) are added to growing polymer chains, one chain ending with an A and the other ending with a B. Two limiting situations may easily be recognized: if the growing chain ends with an A and the probability of adding another A is zero and that of adding a B is 1 (and vice versa for a chain ending with B) a strictly alternating polymer is formed. If, on the other hand, the probability of adding an A is 1 when the chain ends with A and that of adding a B is zero (and again vice versa for a chain ending with B), a mixture of two homopolymers is formed. In both cases the chemical structure of the polymers formed is described in a satisfactory way by an ordinary chemical formula.

In between these two extremes of enzyme specificity, however, an infinite number of distributions can exist, and it is only possible to describe these in a statistical manner. Typical block copolymers will be formed when the probability of adding an A to a chain ending with A is high but different from 1 and the probability of adding a B is small but significantly different from zero.

This simple mechanism can, however, not explain the formation of the three different block types of alginate (Painter *et al.*, 1968). A satisfactory description of the alginate structure is obtained, however, if it is assumed that the specificity of the enzyme is governed not by the ultimate unit of the growing chain but by the penultimate unit (Larsen *et al.*, 1969). Our knowledge of the detailed structure of alginate is not yet sufficient to decide whether a penultimate mechanism alone can explain the monomer sequence of the alginate molecule. The evidence so far indicates that an enzyme specificity governed by the identity of the penultimate unit of a growing polymer chain is the simplest requirement to be imposed upon the enzyme system.

The preceding discussion has been given in terms of a growing polymer chain where the addition of the two types of monomers is governed by transferases. If, on the other hand, the L-guluronic acid residues are formed by an epimerase working at the polymer level, the same reasoning applies and leads to the conclusion that the specificity of the epimerase must involve the recognition of the next nearest neighbour to the unit undergoing epimerization.

Some points of general validity for heteropolymers synthesized by non-template processes may be emphasized. A certain degree of heterogeneity of structure in the sense that all molecules do not have identical sequences, might be expected to be the rule rather than the exception and, in the case of block-copolymers, molecules with different proportions of monomers may occur, depending upon the length of the blocks and the degree of polymerization of the polymer (Haug *et al.*, 1969). A complete description of the structure of such polysaccharides must, therefore, be given in a statistical manner. It is particularly relevant in a discussion of the biosynthesis of polysaccharides to stress that a knowledge of the specificity of the enzyme systems involved may give direct information about the structure of the polymer and, vice versa, that information about the structure of the polymer may tell us about the specificity of the enzyme system.

REFERENCES

Barber, G. A. (1971). *Archs Biochem. Biophys.* **147**, 619.
Barber, G. A. and Chang, M. T. Y. (1967). *Archs Biochem. Biophys.* **118**, 659.
Droyer, M. R., Smydzuk, J. and Smillie, R. M. (1970). *Aust. J. biol. Sci.* **23**, 1005.
Fransson, L.-Å., Malmström, A., Lindahl, U. and Höök, M. (1974). "Biology of the Fibroblast." Academic Press, London and New York.
Fredrick, J. F. (1971). *Physiol. Pl.* **24**, 55.
Glaser, L. and Kornfeld, S. (1961). *J. biol. Chem.* **236**, 1795.
Goldenberg, S. H. and Marechal, L. R. (1963). *Biochim. biophys. Acta* **71**, 743.
Gorin, P. A. J. and Spencer, J. F. T. (1966). *Can. J. Chem.* **44**, 993.

Haug, A. (1974). Chemistry and Biochemistry of Algal Cell Wall Polysaccharides. *In* "Plant Biochemistry" (D. H. Northcote, ed.), MTP International Review of Science, Biochemistry Section. Medical and Technical Publishing Co., Oxford.

Haug, A. and Larsen, B. (1971). *Carbohydr. Res.* **17**, 297.

Haug, A., Larsen, B. and Baardseth, E. (1968). *Proceedings 6th International Seaweed Symposium*, p. 443. Direccion General de la Pesca Maritima, Madrid.

Haug, A., Larsen, B. and Smidsrød, O. (1966). *Acta chem. scand.* **20**, 183.

Haug, A., Larsen, B. and Smidsrød, O. (1967). *Acta chem. scand.* **21**, 691.

Haug, A., Larsen, B., Smidsrød, O. and Painter, T. (1969). *Acta chem. scand.* **23**, 2955.

Hellebust, J. A. and Haug, A. (1972). *Can. J. Bot.* **50**, 177.

Jacobson, B. and Davidson, E. A. (1962). *J. biol. Chem.* **237**, 638.

Kauss, H. and Hassid, W. Z. (1967a). *J. biol. Chem.* **242**, 3449.

Kauss, H. and Hassid, W. Z. (1967b). *J. biol. Chem.* **242**, 1680.

Kauss, H. and Kandler, O. (1962). *Z. Naturf.* **17B**, 858.

Larsen, B. and Haug, A. (1971a). *Carbohydr. Res.* **17**, 287.

Larsen, B. and Haug, A. (1971b). *Carbohydr. Res.* **20**, 225.

Larsen, B. and Haug, A. (1972). *Proceedings 7th International Seaweed Symposium*, p. 491. University of Tokyo Press, Tokyo.

Larsen, B., Painter, T., Haug, A. and Smidsrød, O. (1969). *Acta chem. scand.* **23**, 355.

Lawson, C. J. and Rees, D. A. (1970). *Nature, Lond.* **227**, 392.

Liao, T. H. and Barber, G. A. (1971). *Biochim. biophys. Acta* **230**, 64.

Lin, T. Y. and Hassid, W. Z. (1966a). *J. biol. Chem.* **241**, 3283.

Lin, T. Y. and Hassid, W. Z. (1966b). *J. biol. Chem.* **241**, 5284.

Lindahl, U., Bäckström, G., Malmström, A. and Fransson, L.-Å. (1972). *Biochem. biophys. Res. Commun.* **46**, 985.

Linker, A. and Jones, R. S. (1966). *J. biol. Chem.* **241**, 3845.

Loewus, F. (1965). *Fed. Proc.* **24**, 855.

Loewus, F. (1971). *A. Rev. Pl. Physiol.* **22**, 337.

Loewus, F., Wagner, G., Schiff, J. A. and Weistrop, J. (1971). *Pl. Physiol.* **48**, 373.

Marx-Figini, M. (1969a). *Biochim. biophys. Acta* **177**, 27.

Marx-Figini, M. (1969b). *J. Pol. Sci.* **C28**, 57.

Marx-Figini, M. (1971). *Biochim. biophys. Acta* **237**, 75.

Nagashima, H., Nakamura, S., Nisizawa, K. and Hori, T. (1971). *Pl. Cell Physiol.* **12**, 243.

Nikaido, H. and Hassid, W. Z. (1971). *Adv. Carbohydr. Chem. Biochem.* **26**, 351.

Painter, T. J., Smidsrød, O., Larsen, B. and Haug, A. (1968). *Acta chem. scand.* **22**, 1637.

Pakhomova, I. V. and Zaitzeva, G. N. (1967). *Khim. Biokhim. Uglevodov, Mater. Vses. Konf. 4th*, p. 307.

Percival, E. and McDowell, R. H. (1967). "Chemistry and Enzymology of Marine Algal Polysaccharides." Academic Press, London.

Preiss, J. and Greenberg, E. (1967). *Archs Biochem. Biophys.* **118**, 702.

Ramus, J. and Groves, S. T. (1972). *J. Cell Biol.* **54**, 399.

Rees, D. A. (1961a). *J. chem. Soc.* 5168.

Rees, D. A. (1961b). *Biochem. J.* **81**, 347.

Rees, D. A. (1963). *J. chem. Soc.* 1821.

Robbins, P. W., Bray, D., Dankert, M. and Wright, A. (1967). *Science, N.Y.* **158**, 1536.

Sanwal, G. G. and Preiss, J. (1967). *Archs Biochem. Biophys.* **119**, 454.

Shafizadeh, F. and McGinnis, G. D. (1971). *Adv. Carbohydr. Chem. Biochem.* **26**, 297.

Smidsrød, O. and Haug, A. (1968). *Acta chem. scand.* **22**, 1989.

Smidsrød, O. and Haug, A. (1972). *Acta chem. scand.* **26**, 79.

Stoolmiller, A. C. and Dorfman, A. (1969). *In* "Comprehensive Biochemistry" (M. Florkin and E. H. Stotz, eds), Vol. 17, p. 241. Elsevier, Amsterdam.

Su, J. C. and Hassid, W. Z. (1962). *Biochemistry* **1**, 474.

White, R. C. and Barber, G. A. (1972). *Biochim. biophys. Acta* **264**, 117.

CHAPTER 15

Chemical and Biochemical Aspects of Fungal Cell Walls

R. J. STURGEON

Department of Brewing and Biological Sciences, Heriot-Watt University, Edinburgh, Scotland

I. INTRODUCTION

Chemically the fungal cell wall contains 60–90% polysaccharide material with most of the remainder consisting of protein, lipid, occasionally melanin, polyphosphate and inorganic ions. Physically, the fungal cell wall is a fabric of interwoven microfibrils embedded in amorphous matrix substances. The polysaccharides making up the walls are composed of a variety of sugars. D-Glucose, 2-acetamido-2-deoxy-D-glucose (*N*-acetylglucosamine) and D-mannose are found in most fungi, but the following monosaccharides have been detected in the acid hydrolysates of the fungal cell wall, in amounts varying from traces in some organisms to major components in others: D-galactose, L-fucose, D-xylose, L-rhamnose, L-arabinose, D-glucuronic acid, 2-acetamido-2-deoxy-D-galactose (galactosamine) and D-galacturonic acid (Bartnicki-Garcia, 1968).

The methods commonly used in the study of these polymers include the classical structural methods as used by carbohydrate chemists, but updated by using gas–liquid chromatographic methods combined with mass-spectrometry (Axelsson *et al.*, 1971), enzymic techniques, infrared and X-ray diffraction data (Jones *et al.*, 1972), and phytohaemagglutination procedures. The purpose of this chapter is to describe some of the work on the identification of polymers containing D-glucose and D-glucosamine together with some of the related biochemistry.

II. CHITIN

A. STRUCTURE AND LOCATION

Chitin is a structural component of a large number of fungi, the content varying widely from 1% in baker's yeast cell walls (Bacon *et al.*, 1966) to about 60% in *Allomyces macrogynus* (Aronson and Machlis, 1959) and in *Sclerotium rolfsii* (Bloomfield and Alexander, 1967). The chief difficulty in identifying chitin in the case of *Saccharomyces cerevisiae* was originally caused by the presence of large amounts of apparently insoluble highly branched glucan. Chitin, which is frequently isolated from biological material by removal of contaminating polysaccharides by sequential extractions with alkali, acid and enzymes is a polymer of unbranched chains of β-(1 → 4)-linked 2-acetamido-2-deoxy-D-glucose residues. Although this polymer has been widely reported to be present in the cell walls of mycelium of many fungi, it has also been found in the spore walls of the wheat stem rust (*Puccinia graminis*; Joppien *et al.*, 1972) and in the spore walls of *Pithomyces chartarum* (Sturgeon, 1966). Powder X-ray diffraction still continues to be used as a reliable technique for identifying chitin but in numerous cases the hydrolysis by chitinase to *N,N'*-diacetylchitobiose and *N*-acetyl glucosamine has been used (Berger and Reynolds, 1958; Applegarth and Bozoian, 1969).

Electron microscopic examination of extracted cell ghosts of *Saccharomyces carlsbergensis* has revealed very thin cell envelopes with prominent bud scars in the shape of hollow craters with raised rims. The extracted cell ghosts have a high chitin content which can be removed by digestion with chitinase. This treatment does not destroy the integrity of the cell envelope but largely eliminates the bud-scar rims (Cabib and Bowers, 1971).

B. BIOSYNTHESIS

Particulate preparations from spheroplast lysates of *S. carlsbergensis* are found to catalyse the transfer of 2-acetamido-2-deoxy-D-glucose from UDP-2-acetamido-2-deoxy-D-glucose to an endogenous acceptor. The reaction product can be characterized as chitin by its insolubility in alkali and by the release of *N,N'*-diacetylchitobiose and 2-acetamido-2-deoxy-D-glucose following enzymic hydrolysis. Polyoxin A is a very potent competitive inhibitor of the chitin synthetase, but the antibiotic affects neither the growth of intact cells nor the synthesis of chitin by naked spheroplasts, and this suggests that the chitin synthetase is not located on the outside of the cytoplasmic membrane (Keller and Cabib, 1971). The soluble fraction of a spheroplast lysate of *S. carlsbergensis* and *S. cerevisiae* contains a heat-stable proteinaceous inhibitor of chitin synthetase which is considered to be an allosteric effector serving to regulate the enzyme *in vivo*. The necessity for such regulation is indicated by the fact that chitin is deposited in a very restricted region of the cell wall and probably only during a specific phase of the budding process (Cabib and Keller, 1971).

At levels comparable with those required for the inhibition fungal growth, polyoxin D inhibits the incorporation of 2-amino-2-deoxy-D-[^{14}C]glucose into the cell wall chitin of *Neurospora crassa*. At the same time UDP-2-acetamido-2-deoxy-D-glucose accumulates, indicating inhibition of chitin synthesis. As in the previous example the inhibition is found to be competitive and specific for chitin synthetase (Endo *et al.*, 1970). High resolution autoradiography of *Choanephora curcurbitarum* conidial and hyphal walls, after incorporation of 2-acetamido-2-deoxy-D-[^{3}H]glucose indicates that the incorporation is preferentially into the inner layer of the conidial wall and also in the wall of the growing hyphae (Manocha and Lee, 1972). Similar studies using cell free extracts of *Mucor rouxii* have demonstrated the incorporation of 2-acetamido-2-deoxy-D-[^{14}C]glucose in chitin (McMurrough and Bartnicki-Garcia, 1970). Autoradiographic studies show that the chitin synthetase is localized preferentially in the apical region of the hyphal walls of *M. rouxii* and that this is also the region of chitin deposition *in vivo* (McMurrough *et al.*, 1971). The same labelled compound has been used to follow the sites of chitin incorporation in hyphae of an *Aspergillus nidulans* mutant blocked in amino sugar synthesis and it has been demonstrated that the growing hyphae incorporate the sugar almost exclusively at the tip. The use of cycloheximide causes an increase in the label in subapical regions and reduces that at the tip of the hyphae. At the same time cycloheximide induces morphogenetic changes in the hyphae which produce abnormally large numbers of branches and septa (Katz and Rosenberger, 1971).

III. CHITOSAN

Chitosan is a poorly acetylated or non-acetylated polymer of β-(1 → 4)-linked 2-amino-2-deoxy-D-glucose units which was reported to be present in the cell walls of the mycelium and sporangiophores of *Phycomyces blakesleeanus* (Kreger, 1954). It also occurs in the mycelial, yeast and sporangiophore walls of *Mucor* (Bartnicki-Garcia and Nickerson, 1962; Bartnicki-Garcia and Reyes, 1965). Although chitin is also normally present in these walls, but at a lower level than the chitosan, glucose-containing polymers are normally absent and it has been speculated that chitosan fulfils the structural role played by glucan in other fungi (Bartnicki-Garcia, 1968). However, the spore walls of *Puccinia graminis* appear to contain chitosan together with a galactoglucomannan: in contrast the germ tube wall contains chitin (Joppien *et al.*, 1972).

IV. CARBOHYDRATE–PROTEIN POLYMERS

In the isolated cell walls of baker's yeast Northcote and Horne (1952) have demonstrated the presence of two polysaccharides, a glucan and a mannan, which are associated with proteinaceous material. Subsequently Falcone and Nickerson (1956) reported the isolation of a homogeneous structural entity consisting of protein and mannan in the ratio of 1:12. This was followed by the isolation of a glucan–protein complex and two gluco-mannan–protein complexes from *Saccharomyces cerevisiae* (Kessler and Nickerson, 1959). Using a different extraction procedure two glycoprotein

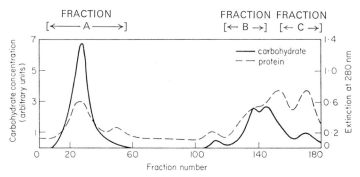

FIG. 1. DEAE-cellulose chromatography of glycoproteins from *Pithomyces chartarum* using a NaCl gradient (0–0.5 M) at pH 7.0.

fractions have been isolated from baker's yeast in which mannose is the major carbohydrate component in one and glucose and mannose in the other. Each of the complexes contains 2-amino-2-deoxy-D-glucose and the former glycoprotein appears to be homogeneous by the criterion of analytical ultracentrifugation and moving boundary electrophoresis (Korn and North-cote, 1960). However, not all glycoproteins which have been extracted from fungal sources have proved to be homogeneous. A series of glycoproteins have been isolated from *Pithomyces chartarum* (Sturgeon, 1964). One of the extracted fractions has been submitted to ion-exchange chromatography (Jack, 1971) and it has been shown to contain at least five components (Fig. 1). Analysis of the individual components has shown that they differ considerably in the amount of hexose, protein and amino sugars (Table I). One of the unusual features is the apparent absence of hexosamine in two of the glyco-protein fractions.

TABLE I

Composition of purified glycoproteins from *P. chartarum*[a]

Fraction	Carbohydrate (%)	Protein (%)	Hexosamine (%)
A	75	14	11
B	58	42	0
C	40	60	0

[a] Jack (1971). See Fig. 1.

A. CARBOHYDRATE–PROTEIN LINKAGE

Fungal glycoproteins, on hydrolysis, yield 15 or 16 amino acids, as commonly found in plant and animal glycoproteins. This contrasts with the cell walls of gram-positive bacteria which contain relatively few constituent amino acids.

The nature of the linkage between the carbohydrate and protein parts of the glycoprotein molecules has been investigated by numerous workers and the existence of three forms has been demonstrated. One of the earliest suggestions regarding the type of linkage was made by Kessler and Nickerson (1959) who observed the high proportion of acidic amino acid residues in a glycoprotein of *Saccharomyces cerevisiae* and suggested that there was an ester linkage between a hydroxyl group on the carbohydrate and a carboxyl group on an acidic amino acid. The existence of an alternative type of linkage was suggested by Korn and Northcote (1960) as a result of their studies on the glycoproteins isolated from baker's yeast. All the carbohydrate–protein complexes which they isolated contained glucosamine and they speculated that this was the bridging sugar between the carbohydrate and protein moieties. It was later concluded that high molecular weight mannan of this polymer was joined to the protein portion of the glycoprotein by an *N*-(β-aspartyl)-β-D-(*N*-acetyl)glucosaminide linkage (Sentandreu and Northcote, 1968). (Table II.)

Carbohydrate–protein linkages involving *O*-glucosidic linkages to serine and threonine were first demonstrated in mammalian mucopolysaccharides (Anderson *et al.*, 1963) but they have since been found to be widespread in nature. Sentandreu and Northcote (1968) were the first to detect their presence in fungal glycoproteins. The same mannan–protein complex of *S. cerevisiae* which had been shown to contain hexosamine–aspartamide linkage, can be digested with proteolytic enzymes to produce the corresponding glycopeptide. The peptide component, which is rich in threonine and serine, undergoes β-elimination in dilute alkali releasing mannose and mannose-containing oligosaccharides. Amino acid analysis of the acid hydrolysed product shows a considerable loss of serine and threonine compared with the levels of those

TABLE II

Types of carbohydrate–protein linkages reported to occur in fungal cell walls

2-acetamido-1-(L-β-aspartamido)-
1,2-dideoxy-D-glucosyl

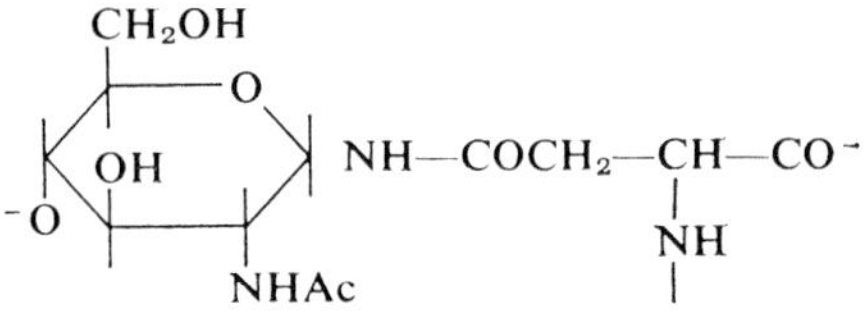

$$CH_2OH$$

(structure of glucosyl ring with OH, NHAc) NH—COCH$_2$—CH—CO$^-$
                                              |
                                             NH
                                              |

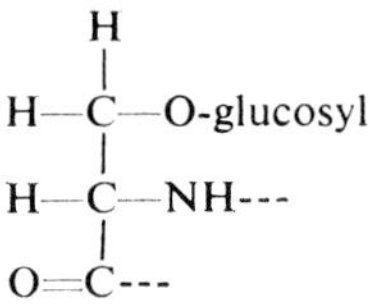

O-glucosylseryl

$$
\begin{array}{c}
H \\
| \\
H—C—O\text{-glucosyl} \\
| \\
H—C—NH\text{---} \\
| \\
O{=}C\text{---}
\end{array}
$$

O-Glycosylthreonyl

$$
\begin{array}{c}
CH_3 \\
| \\
H—C—O\text{-Glycosyl} \\
| \\
H—C—NH\text{---} \\
| \\
O{=}C\text{---}
\end{array}
$$

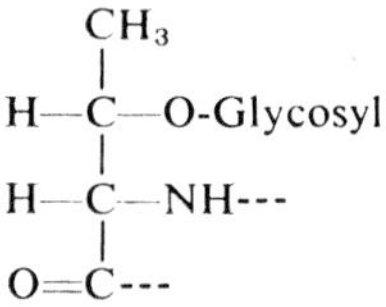

amino acids in the original glycopeptide. This is consistent with other evidence of the cleavage of mannosyl-O-seryl and mannosyl-O-threonyl linkages in alkaline conditions (Fig. 2). Thus in the case of the *S. cerevisiae* mannan–protein, the hydroxyamino acids appear to be linked to low molecular weight carbohydrate units and the aspartic acid to high weight mannan.

Treatment of the glycopeptide of *Pithomyces chartarum* with NaOH is accompanied by the release of most of the protein moiety and a decrease in the serine and threonine content (Table III). The results suggest that the carbohydrate, in this case, is present in a high molecular weight form and linked to the protein moiety by O-glycosidic bonds to the hydroxyamino acids (Jack, 1971).

A galactomannan from the yeast form of *Cladosporium werneckii* has been shown to contain phosphate groups (3–5%) and covalently linked peptide (10%) (Lloyd, 1970). Treatment of the complex with 0·3 M NaOH and 0·3 M NaBH$_4$ cleaves the carbohydrate–protein linkages, with destruction of serine and threonine, and concomitant release of two types of carbohydrate

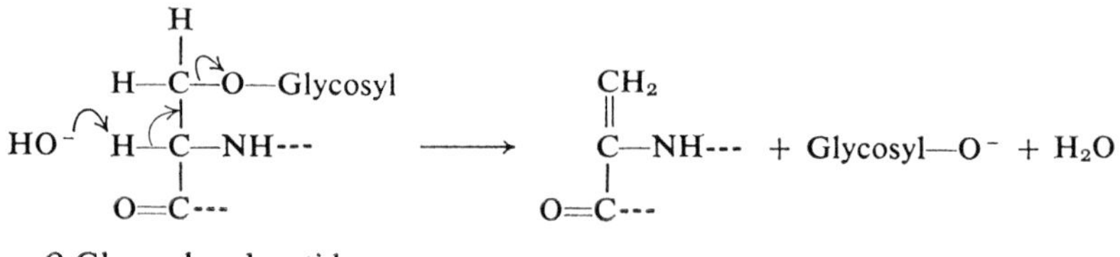

O-Glycosylserylpeptide

FIG. 2. β-Elimination of the O-glycosylserylpeptide linkage in dilute alkali.

TABLE III

Effect of alkali on the amino acid compositions[a] of the peptidogalactomannans from *C. werneckii* [A] and *P. chartarum* [B][b]

Amino acid	A		B	
	Original	After 0·3 M NaOH + 0·3 M NaBH₄	Original	After 0·5 M NaOH
Asp	64	63	54	63
Thr	248	48	143	113
Ser	153	66	173	112
Glu	81	81	88	105
Pro	51	43	13	13
Gly	65	93	114	129
Ala	120	127	137	168
Val	45	39	111	101
Ile	31	29	32	27
Leu	34	44	36	34
Tyr	Tr	9	23	48
Phe	20	24	14	34
Lys	16	21	14	15
His	4	7	25	22
Arg	7	12	10	13

[a] nmol/mg
[b] Jack (1971)

chains: (a) small, reduced mannose-containing oligosaccharides; and (b) large phosphorylated galactomannans with a molecular weight of about 60 000. The amino acid analyses are given in Table III. From the results of partial acid hydrolysis experiments and other evidence it has been proposed that the general structure of the peptide-phosphogalactomannan complex is as represented in Fig. 3.

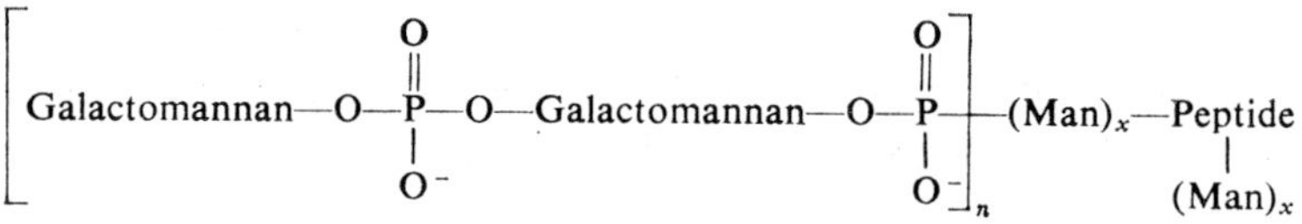

$$\left[\text{Galactomannan} - O - \overset{\overset{O}{\|}}{\underset{\underset{O^-}{|}}{P}} - O - \text{Galactomannan} - O - \overset{\overset{O}{\|}}{\underset{\underset{O^-}{|}}{P}} \right]_n - (\text{Man})_x - \underset{(\text{Man})_x}{\overset{|}{\text{Peptide}}}$$

FIG. 3. Galactomannan-peptide from *Cladosporium werneckii* (Lloyd, 1970).

V. MISCELLANEOUS COMPLEX POLYSACCHARIDES

The earlier discussion has shown that amino sugars in fungal polymers are normally found in the form of homopolysaccharides, such as chitin or chitosan, or are involved in the carbohydrate–protein linkage of glycoproteins. However, a number of exceptions have been reported.

Cells of *Pichia bovis* and *Saccharomyces phaseolosporus* have been extracted with alkali and the solubilized polysaccharides purified by copper complexing. The polymers, which are monodisperse in aqueous solutions, giving single peaks on ultracentrifugation, contain predominantly D-mannose as well as 2-acetamido-2-deoxy-D-glucose and 6% protein. Methylation analysis has indicated that the amino sugar occurs as non-reducing end groups on the polysaccharide chain. The location of the amino sugar thus differs from that in the mannan-protein complex in which 2-acetamido-2-deoxy-D-glucose residues serve as a bridge between the mannan and protein (Gorin *et al.*, 1971).

A further example of the unusual role of amino sugars in fungal cell wall polymers has been shown in the characterization of a yeast mannan containing 2-acetamido-2-deoxy-D-glucose as an immunochemical determinant (Raschke and Ballou, 1972). This mannan from *Kluyveromyces lactis* is characterized by a high content of the amino sugar, most of which is located in penta-saccharide side chains containing four mannose units and one 2-acetamido-2-deoxy-D-glucose unit. This side chain is released from the mannan by con-

$$\alpha\text{-Man-}(1 \rightarrow 3)\text{-}\alpha\text{-Man-}(1 \rightarrow 2)\text{-}\alpha\text{-Man-}(1 \rightarrow 2)\text{-}\alpha\text{-Man}$$

$$\uparrow 2$$

$$\mid 1$$

$$\alpha\text{-GlcNAC}$$

FIG. 4. Pentasaccharide side chains of *Kluyveromyces lactis* (Raschke and Ballou, 1972).

trolled acetolysis, along with mannose, mannobiose, mannotriose, manno-tetraose and a fragment containing 1 mol of amino sugar and 3 of mannose. The structure of the pentasaccharide side chain can be shown by methylation analysis to consist of a mannotetraose unit with an α-linked 2-acetamido-2-deoxy-D-glucose residue attached to position 2 of the penultimate mannose in the chain (Fig. 4). The presence of the amino sugar protects this type of side chain from digestion by α-mannanase. Rabbit antiserum formed against intact *K. lactis* cells has two specificities, one for the mixed sugar penta-saccharide and one for mannotetraose side chains.

VI. GLUCANS

The feature of fungal polysaccharides that is most striking is the relatively large variation in the types of glycosidic linkages present, namely $(1 \rightarrow 2)$, $(1 \rightarrow 3)$, $(1 \rightarrow 4)$, and $(1 \rightarrow 6)$. Many of the true fungal polysaccharides are homopolymers containing only one monosaccharide unit, most commonly glucose. In a large number of cases this sugar is β-linked in the polymer although in a few instances α-linkages are to be found (Table IV).

TABLE IV

Cell wall glucans

Fungus	Linkage	Reference
Poria cocos	β-(1 → 3)	Warsi and Whelan (1957)
Verticillium alboatrum	α-(1 → 3)-, β-(1 → 4)-, β-(1 → 6)-	Wang and Bart-nicki-Garcia (1970)
Aspergillus nidulans	β-(1 → 3)-, β-(1 → 6)-, α-(1 → 3)- α-(1 → 4)-	Bull (1970)
Aspergillus niger	α-(1 → 3)-, α-(1 → 4)-	Barker *et al.* (1952)
Fusiococcum amygdali	β-(1 → 3)-, β-(1 → 6)-, α-(1 → 4)-	Buck and Obaidah (1971)
Pithomyces chartarum	β-(1 → 3)-, β-(1 → 6)-, α-(1 → 4)-	Jack (1971)

A. GLUCANS CONTAINING β-(1 → 3)- AND β-(1 → 6)-LINKAGES

Extraction of whole cells of *Saccharomyces cerevisiae* have been used by many workers as the starting point for a structural investigation of β-glucan. Using a mild extraction technique, Zechmeister and Toth (1934) isolated a glucan which on methylation and hydrolysis gave 2,4,6-tri-*O*-methyl-D-glucose as the major product, thus establishing the presence of (1→3)-glucosidic linkages. Confirmation of the existence of β-(1 → 3)-linkages was obtained by the isolation of laminaribiose from the partial acid hydrolysate of the β-glucan (Barry and Dillon, 1943). The presence of β-(1 → 6)-linkages in the yeast glucan was first reported by Peat *et al.* (1958) who isolated gentiobiose, gentiotriose and gentiotetraose, leading to the proposal that the polysaccharide had a linear structure containing certain sequences of β-(1 → 3)- and β-(1 → 6)-linked D-glucose residues. This structure differs considerably from an earlier suggestion of Bell and Northcote (1950), based on methylation analysis, that the molecule was highly branched with β-(1 → 3)-linked glucose residues in side chains linked through β-(1 → 2)-inter chain linkages to a β-(1 → 3)-linked glucan backbone. On the basis of methylation, periodate oxidation and enzymic degradation studies, a significantly different structure was proposed by Manners and Patterson (1966). They regarded the glucan as a branched polysaccharide consisting of main chains of β-(1 → 6)-linked D-glucose residues to which linear side chains of β-(1 → 3)-linked residues were attached, since hydrolysis of the glucan by a bacterial laminarinase released glucose, laminaribiose, laminaritriose and a water-soluble 'limit dextrin' which was resistant to further hydrolysis. The dextrin was shown to have a main chain of β-(1 → 6)-linked glucose residues with short chains of β-(1 → 3)-linked glucose residues attached to the main chain at C-3. A structure of the same general type, but differing in some details, such as the average chain length, was proposed independently by Misaki *et al.* (1968). Infrared spectroscopic and other evidence led Bacon and Farmer (1968) to suggest that these different interpretations of structure arise, in part, from the heterogeneous nature of the yeast glucan, due to the presence of at least two polysaccharides, one of which is a β-(1 → 6)-glucan. This report has been confirmed and further chemical evidence on the structures of the two glucans has been presented (Manners and Masson, 1969) on the basis of the following evidence.

Exhaustive extraction of the alkali-insoluble residue from the yeast using 0·5 M acetic acid at 90°C solubilizes glycogen and a glucan which on partial acid hydrolysis gives rise to glucose and gentiosaccharides. After removal of the contaminating glycogen, the glucan, which has a degree of polymerization of 140 ± 10, can be readily attacked by a purified endo-β-(1 → 6)-glucanase liberating a series of gentiosaccharides, but cannot be hydrolysed by bacterial or fungal β-(1 → 3)-glucanase. When taken in conjunction with periodate oxidation and Smith degradation procedures (periodate oxidation, borohydride reduction and mild acid hydrolysis) the evidence points to the presence of a high proportion of β-(1 → 6)-linked glucose residues, together with a small number of β-(1 → 3)-linked residues.

It is clear that the various structures proposed by previous workers are due to the presence of differing amounts of the β-(1 → 6)-linked glucan in preparations of the β-(1 → 3)-glucan. The latter appears to be a large insoluble molecule containing about 1500 glucose residues with a low degree of branching.

Pachyman, the β-glucan, which can be isolated from *Poria cocos* Wolf, has been known for many years in Japan and China as the traditional cure-all, Bukuryo. The fungus, which grows around pine roots, contains about 92% polyglucose after defatting. Partial acid hydrolysis yields a homologous series of β-(1 → 3)-linked glucose oligosaccharides up to laminaripentaose, and on the basis of this and methylation data, Warsi and Whelan (1957) concluded that the molecule was a linear β-(1 → 3)-linked glucan. When the structure of the polysaccharide was investigated by Saito *et al.* (1963) it was reported that the molecule contained a few β-(1 → 6)-linked glucose residues and that it could possess four branch points and had a degree of polymerization (DP) of 255. The existence of the β-(1 → 6)-linkages has been confirmed (Chihara *et al.*, 1970) and Smith degradation furnished a linear glucan (with antitumour activity), that was not present in the original polysaccharide. Carboxymethylpachyman, labelled at the carboxymethyl residue with ^{14}C, has been prepared as a water soluble derivative with strong antitumour activity so that its distribution at cellular and subcellular levels can be investigated (Hamuro *et al.*, 1971). A further minor revision of the structure of the native pachyman molecule has been suggested (Hoffman *et al.*, 1971) in which there are about 700 β-(1 → 3)-linked glucose residues with approximately six branch points (possibly at C-2) together with a few (1 → 6)-linked internal residues which appear to be branched at C-4.

B. GLUCANS CONTAINING β-(1 → 4)-LINKAGES

Phytophthora, a notorious genus of plant pathogens, is a prominent example of one of the relatively few groups of fungi with cellulosic walls. The presence of cellulose in *Phytophthora* has been established by X-ray analysis by Frey (1950). Qualitative evidence for the presence of cellulose in the hyphal wall of *P. cinnamomi* and *P. parasitica* includes the release of about 20% of cellobiose during enzymic digestion of *P. cinnamomi* walls with a crude cellulase. The walls had a very limited solubility in strong alkali but were partially soluble in Schweitzer's reagent and a polymer with an X-ray pattern similar to that of cellulose could subsequently be regenerated (Bartnicki-Garcia, 1966).

After removal of the cellulose from the wall of *P. cinnamomi* with Schweitzer's reagent a residual glucan is left: methylation studies and periodate oxidation, followed by reduction and hydrolysis (which affords glycerol, erythritol and glucose in a molar ratio of 23:10:66), show that the residual glucan is a highly branched polysaccharide composed predominantly of β-(1 → 3)-linked glucose chains with branch points at C-3 and C-6. The β-(1 → 3)-linked chains are probably very short with an average of 4–5 units. In addition, about 10% of the glucose units appear to be joined by (1 → 4)-linkages, possibly corresponding to cellulose chains firmly bound to the non-cellulosic portion of the insoluble hyphal wall glucan (Zevenhuizen and Bartnicki-Garcia, 1969).

Although cellobiose has been isolated from the enzymic digestion of the alkali insoluble glucan of *Verticillium albo-atrum*, the polymeric nature of the β-(1 → 4)-linked glucose units remains uncertain (Wang and Bartnicki-Garcia, 1970). The microfibrillar network consists mainly of β-(1 → 3)-glucan and chitin.

C. GLUCANS WITH α-(1 → 3)- AND α-(1 → 4)-LINKAGES

The cell walls of *Aspergillus niger* have been shown to contain a glucan, nigeran, with high optical rotation, $[\alpha]_D + 283°$. From the controlled acid hydrolysis of nigeran, two oligosaccharides, α-D-glucopyranosyl-(1 → 3)-D-glucose (Barker *et al.*, 1952; 1953) and α-D-glucopyranosyl-(1 → 3)-α-D-glucopyranosyl-(1 → 4)-D-glucose (Barker *et al.*, 1957) have been isolated. Methylation analysis shows the presence of approximately equal numbers of (1 → 3)- and (1 → 4)-linked residues, and all evidence indicates that nigeran is a linear glucan in which α-(1 → 3)- links alternate with α-(1 → 4)-links. A more recent investigation of the *A. niger* cell wall has revealed that after removal of the nigeran from the mycelium, another highly dextrorotatory glucan can be isolated (Johnston, 1965a) which, on partial acid hydrolysis, yields nigerose and a homologous series of α-(1 → 3)-linked oligosaccharides (Johnston, 1965b). Methylation analysis, periodate oxidation and Smith degradation of this pseudonigeran have indicated that the polymer has a

DP of 330 and contains mainly α-(1 → 3)-linkages (Horisberger *et al.*, 1972). About 1·4% of (1 → 4)-linkages are present, either in the pseudonigeran or arising from contaminating material. However, the periodate uptake by the polysaccharide and formic acid produced is somewhat higher than would be expected for a linear (1 → 4)-glucan. On this basis and from the results of acid hydrolysis investigations, it has been suggested that there may be a very small percentage of (1 → 4)-linkages present near the reducing end of the molecule (Hasegawa *et al.*, 1969).

A similar α-glucan has been found in the cell walls of *Fusiococcum amygdali* (Obaidah and Buck, 1971). After treatment with α-amylase, which only has a slight action on the molecule, the product, which comprises 70% of the original glucan, can be hydrolysed to produce a homologous series of nigerodextrins, indicating a predominance of α-(1 → 3)-linkages. Maltose can also be detected, showing that a small number of α-(1 → 4)-linked glucopyranose residues are present, and methylation analysis indicates that the polymer is a linear glucan with an average chain length of 230 residues and is composed of 97·6% (1 → 3)-linkages, the remainder being (1 → 4)- and/or (1 → 6)-linkages. Long blocks of (1 → 4)-linkages are considered unlikely and the lowering of the average molecular weight and increase in polydispersity after periodate oxidation is taken to indicate that most of the α-(1 → 4)-linkages are arranged at internal positions in the chain and not all at the reducing end, as has been suggested for pseudonigeran (Hasegawa *et al.*, 1969).

D. GLUCANS CONTAINING BOTH α- AND β-LINKED RESIDUES

The cell walls of *Fusiococcum amygdali*, an organism which causes the wilting of almond and peach trees, has been shown to consist of 85% poly-saccharide material (Buck and Obaidah, 1971) of which D-glucose is the main constituent. Cell walls of the fungus can be stained dark blue with iodine and they are attacked by α-amylase, with the liberation of glucose, maltose and maltotriose, indicating the existence of chains of α-(1 → 4)-linked glucopyranose residues. Glucose and gentiobiose are liberated from the cell walls by the action of an exo-β-(1 → 3)-glucanase, giving evidence for both β-(1 → 3)- and β-(1 → 6)-glucopyranose linkages. Fractionation of the cell wall complex by alkali releases three principal glucan fractions, one of which contains predominantly β-linkages and one which contains both α- and β-linkages. This latter component which possesses both α-(1 → 4)- and β-(1 → 3)-linkages, and hence can be hydrolysed by both α-amylase and exo-β-(1 → 3)-glucanase, sediments in the ultracentrifuge as a single boundary ($s_{20}^{\circ}6S$). Periodate oxidation, methylation and other studies suggest that this 6S polysaccharide consists of a branched galactomannorhamnan core to which are attached chains of α-(1 → 4)-linked glucopyranose residues and a branched β-(1 → 3)-, β-(1 → 6)-glucan.

Our investigations (Jack and Sturgeon, unpublished results) on part of the cell wall of *Pithomyces chartarum* have shown that it has a number of similarities to the *F. amygdali* cell wall. When a hyphal wall fraction is exhaustively extracted with the sequence of 10% trichloracetic acid, anhydrous ethylenediamine and aqueous sodium hydroxide, about 45% of the original cell wall remains insoluble and this contains only D-glucose (80%) and 2-acetamido-2-deoxy-D-glucose (20%). The amino sugar is almost certainly present as chitin, since it can be completely digested by a chitinase from *Streptomyces albidoflavus*. Infrared spectroscopy has shown the presence of both α- and β-linkages in the glucan which is isolated after chitinase treatment. Periodate oxidation followed by borohydride reduction and acid hydrolysis releases glucose (70%), erythritol and glycerol, which suggests that a large part of the molecule consists of $(1 \rightarrow 3)$-linked glucose units and that $(1 \rightarrow 4)$-linkages are also present which give rise to the erythritol. This is substantiated by the isolation, after mild hydrolysis, of glucosylerythritol and a homologous series of laminarisaccharides, each terminated with erythritol. The results of mild acid hydrolysis of the original glucan, using polystyrene sulphonic acid (Painter, 1960), and also of digestion of the glucan with enzymes, confirm that the erythritol is derived from *O*-glucosyl residues in the molecule. In both cases, a homologous series of laminarisaccharides, and maltose, maltotriose and maltotetraose are released. Methylation results substantiate these gross structural features and in addition to indicating the presence of $(1 \rightarrow 6)$-linkages also point to a fairly high degree of branching. In many ways, this glucan is similar to the non-cellulosic glucan, isolated from *Phytophthora cinnamomi* (Zevenhuizen and Bartnicki-Garcia, 1969) which also contains approximately 70% $(1 \rightarrow 3)$- in addition to $(1 \rightarrow 4)$ and $(1 \rightarrow 6)$-linkages, although all have the β-conformation. Smith degradation of the *P. cinnamomi* glucan does not yield glucosylerythritol or the erythritol glucosides of laminarisaccharides inferring that there are no random or alternating $(1 \rightarrow 3)$-bonds, which is the case for the glucan from *P. chartarum*.

REFERENCES

Anderson, B., Hoffman, K. and Meyer, K. (1963). *Biochim. biophys. Acta* **74**, 309.

Applegarth, D. A. and Bozoian, G. (1969). *Archs Biochem. Biophys.* **134**, 285.

Aronson, J. M. and Machlis, L. (1959). *Am. J. Bot.* **46**, 292.

Axelsson, K., Bjorndal, H., Svensson, S. and Hammarstrom, S. (1971). *Acta chem. scand.* **25**, 3645.

Bacon, J. S. D., Davidson, E. D., Jones, D. and Taylor, I. F. (1966). *Biochem. J.* **101**, 36c.

Bacon, J. S. D. and Farmer, V. C. (1968). *Biochem. J.* **110**, 34P.

Barker, S. A., Bourne, E. J. and Stacey, M. (1952). *Chemy Ind.* 576.

Barker, S. A., Bourne, E. J. and Stacey, M. (1953). *J. chem. Soc.* 3084.

Barker, S. A., Bourne, E. J., O'Mant, D. M. and Stacey, M. (1957). *J. chem. Soc.* 2248.

Barry, V. C. and Dillon, T. (1943). *Proc. R. Irish Acad.* **49B**, 177.

Bartnicki-Garcia, S. (1966). *J. gen. Microbiol.* **42**, 57.

Bartnicki-Garcia, S. (1968). *A. Rev. Microbiol.* **22**, 87.

Bartnicki-Garcia, S. and Nickerson, W. J. (1962). *Biochim. biophys. Acta* **58**, 102.

Bartnicki-Garcia, S. and Reyes, E. (1965). *Bact. Proc.* 26.

Bell, D. and Northcote, D. H. (1950). *J. chem. Soc.* 1944.

Berger, L. R. and Reynolds, D. M. (1958). *Biochim. biophys. Acta* **29**, 522.

Bloomfield, B. J. and Alexander, M. (1967). *J. Bact.* **93**, 1276.

Buck, K. W. and Obaidah, M. A. (1971). *Biochem. J.* **125**, 461.

Bull, A. T. (1970). *J. gen. Microbiol.* **63**, 75.

Cabib, E. and Bowers, B. (1971). *J. biol. Chem.* **246**, 152.

Cabib, E. and Keller, F. A. (1971). *J. biol. Chem.* **246**, 167.

Chihara, C., Hamuro, J., Maeda, Y., Avai, Y. and Fukuoka, F. (1970). *Nature, Lond.* **225**, 943.

Endo, A., Kakiki, K. and Misato, T. (1970). *J. Bact.* **104**, 189.

Falcone, G. and Nickerson, W. J. (1956). *Science, N.Y.* **124**, 272.

Frey, R. (1950). *Ber. Schweiz. Bot. Ges.* **60**, 199.

Gorin, P. A. J., Spencer, J. F. T. and Finlayson, A. J. (1971). *Carbohydr. Res.* **16**, 161.

Hamuro, J., Yamashita, Y., Ohsaka, Y., Maeda, Y. and Chigara, G. (1971). *Nature, Lond.* **233**, 486.

Hasegawa, S., Nordin, J. H. and Kirkwood, S. (1969). *J. biol. Chem.* **244**, 5460.

Hoffman, G. C., Simson, B. W. and Timell, T. E. (1971). *Carbohydr. Res.* **20**, 185.

Horisberger, M., Lewis, B. A. and Smith, F. (1972). *Carbohydr. Res.* **23**, 183.

Jack, W. (1971). Ph.D. Thesis, Heriot-Watt University.

Johnston, I. R. (1965a). *Biochem. J.* **96**, 651.

Johnston, I. R. (1965b). *Biochem. J.* **96**, 659.

Jones, D., Farmer, V. C., Bacon, J. S. D. and Wilson, M. (1972). *Trans. brit. mycol. Soc.* **59**, 11.

Joppien, J., Burger, A. and Resisener, H. J. (1972). *Arch. Mikrobiol.* **82**, 373.

Katz, D. and Rosenberger, R. F. (1971). *J. Bact.* **108**, 184.

Keller, F. A. and Cabib, E. (1971). *J. biol. Chem.* **246**, 100.

Kessler, G. and Nickerson, W. J. (1959). *J. biol. Chem.* **234**, 2281.

Korn, E. and Northcote, D. H. (1960). *Biochem. J.* **75**, 12.

Kreger, D. R. (1954). *Biochim. biophys. Acta* **13**, 1.

Lloyd, K. O. (1970). *Biochemistry* **9**, 3446.

Manners, D. J. and Masson, A. J. (1969). *FEBS Lett.* **4**, 122.

Manners, D. J. and Patterson, J. C. (1966). *Biochem. J.* **98**, 19C.

McMurrough, I. and Bartnicki-Garcia, S. (1970). *Biochem. J.* **98**, 19C.

McMurrough, I., Flores Carreon, A. and Bartnicki-Garcia, S. (1971). *J. biol. Chem.* **246**, 3999.

Manocha, M. S. and Lee, K. Y. (1972). *Can. J. Bot.* **50**, 35.

Misaki, A., Johnston, J., Kirkwood, S., Scaletti, J. V. and Smith, F. (1968). *Carbohydr. Res.* **6**, 150.

Northcote, D. H. and Horne, R. W. (1952). *Biochem. J.* **51**, 232.

Obaidah, M. A. and Buck, K. W. (1971). *Biochem. J.* **125**, 473.

Painter, T. J. (1960). *Chemy Ind.* 1214.

Peat, S., Whelan, W. J. and Edwards, T. E. (1958). *J. chem. Soc.* 3862.

Raschke, W. C. and Ballou, C. E. (1972). *Biochemistry* **11**, 3807.

Saito, H., Misaki, A. and Harada, T. (1963). *Agric. biol. Chem. (Japan)* **32**, 1261.

Sentandreu, R. and Northcote, D. H. (1968). *Biochem. J.* **109**, 419.
Spiro, R. G. (1970). *A. Rev. Biochem.* **39**, 599.
Sturgeon, R. J. (1964). *Biochem. J.* **92**, 60P.
Sturgeon, R. J. (1966). *Nature, Lond.* **209**, 204.
Wang, M. C. and Bartnicki-Garcia, S. (1970). *J. gen. Microbiol.* **64**, 41.
Warsi, S. A. and Whelan, W. J. (1957). *Chemy Ind.* 1573.
Zechmeister, L. and Toth, G. (1934). *Biochem. Z.* **270**, 309.
Zevenhuizen, L. P. T. M. and Bartnicki-Garcia, S. (1969). *Biochemistry* **8**, 1496.

CHAPTER 16

Glycoproteins of Higher Plants*

NATHAN SHARON

*Department of Biophysics, The Weizmann Institute of Science,
Rehovoth, Israel*

I. INTRODUCTION

Modern biochemical research has taught us that underlying the enormous diversity of living organisms there is a remarkable degree of unity. This has been clearly stated by Baldwin (1949):

> We arrive from several different lines of approach at essentially the same conclusion: there exists a common fundamental ground plan of composition and metabolism to which all animals, and very probably other living organisms also, conform, and that superposed on this foundation there are numerous secondary, specific and adaptational variations, some of addition and others of omission.

At the molecular level this common chemical ground plan is manifested by the fact that all living organisms contain substances composed of a small

* This research was supported in part by a contribution from an anonymous friend of The Weizmann Institute in Buenos Aires, Argentina.

number of monomeric molecules, joined in a limited variety of linkages. Accordingly, it is not at all surprising that glycoproteins, which are widely distributed in the animal kingdom, are also present in plants and that plant glycoproteins exhibit structural features common with those of their animal counterparts, although some differences also exist.

II. General Aspects

Glycoproteins are usually defined (Marshall and Neuberger, 1968, 1970) as macromolecules that are composed of a polypeptide backbone to which are attached one or more carbohydrate moieties. The latter usually consist of 2–15 monosaccharide residues that are composed of two or more of the sugars: D-mannose, D-galactose, D-glucose, L-fucose, N-acetyl-D-glucosamine, N-acetyl-D-galactosamine and sialic acids. In most glycoproteins, the absolute configuration of the sugars has not been proven conclusively. The carbohydrate moieties do not seem to contain continuous sequences of repeating units of the type ···ABAB··· (except in the mucopolysaccharides of connective tissue which are sometimes classified with the glycoproteins), although it is becoming clear that certain regularities of structure do exist.

The earliest suggestion that proteins may indeed contain covalently linked carbohydrates dates back some 120 or 130 years (Neuberger, 1971). In the second half of the nineteenth century it became fairly well established that the highly viscous proteins known as mucus or mucins, found in external secretions such as saliva, sweat and the covering layer of the digestive and respiratory tracts of animals, are glycoproteins. Most ordinary proteins were, however, believed to be free of sugars. With the advances in our knowledge of the properties of proteins and the great improvements in the techniques for their study it has become evident during the last 20 years that glycoproteins are of very wide occurrence in nature. At present is seems that the majority of proteins are glycoproteins.

Because of the intense interest in the biology of higher animals and of mammals in particular, glycoproteins derived from secreted fluids or circulating body fluids were investigated in great detail. However, the study of the biochemistry of plant tissues, and in particular of plant proteins, was neglected for many years, and the occurrence of glycoproteins in these sources was established at a relatively late date. In fact, it is only some 10 years since the first unequivocal evidence for the occurrence of glycoproteins in higher plants was provided. This delay may have been caused by the incorrect belief that amino sugars, which are characteristic constituents of glycoproteins, were as a rule absent from plant tissues (Sharon, 1965). It is now known that D-glucosamine is a widespread component of plants, although a relatively minor one. Other amino sugars are very rarely found in plants, if at all.

Plant glycoproteins are not confined to a single species or to one part of the plant, but appear to be widely distributed both phylogenetically and anatom-

TABLE I

Distribution and biological activities of plant
glycoproteins

Lectins (phytoagglutinins)[a]
  soybean agglutinin, wax bean agglutinin
Enzymes
  pineapple bromelain, horseradish peroxidases
Toxins and allergens
  ricin (castor bean), pollen allergens (rye grass, ragweed)
Structural (cell wall) proteins
  extensin (tomato, carrot, sycamore)
Others
  7*S* protein of soybean
  glycoproteins I and II of kidney bean

---

[a] For complete list, see Table IV.

ically. They include lectins (phytoagglutinins), enzymes, toxins, structural proteins and other proteins for which no biological activity or function has been found (Table I). Although the role of the saccharide moieties in some animal glycoproteins is beginning to be understood, nothing is known about their function in plant glycoproteins.

The isolation of plant glycoproteins and their characterization with respect to their chemical structure, as well as their physical and biological properties are essential to the understanding of the role of these compounds in plants. Glycoproteins, whether from plants or from other sources, are generally isolated and purified by the usual techniques of protein chemistry, with every effort being made to obtain the materials in their native undenatured state. The presence of carbohydrate offers an additional handle to monitor fractionations, and to assess the purity and homogeneity of the final product. It is also responsible for the interaction of glycoproteins with lectins (Sharon and Lis, 1972; Lis and Sharon, 1973). The ability of lectins to bind sugars is now being used for the development of specific affinity chromatography methods for the isolation of glycoproteins. For example, immobilized concanavalin A has been used for the isolation of immunoglobulins and of glycoprotein enzymes, such as glucose oxidase (see Sharon and Lis, 1972). Although no reports of the use of lectins for the isolation of plant glycoproteins have appeared, there is no doubt that this new tool will also play an important role in the study of glycoproteins from plants.

For unequivocal demonstration that a protein is indeed a glycoprotein, it is necessary to isolate from its digests a low molecular weight glycopeptide containing both amino acid and carbohydrate residues. Such glycopeptides have been isolated from only a very small number of plant glycoproteins. However, for many other glycoproteins mentioned in this chapter there is satisfactory evidence of their homogeneity and so it may be safely assumed that the carbohydrate is indeed an integral part of the molecule, covalently linked to the polypeptide chain.

The carbohydrate content of plant glycoproteins appears to cover a narrower range than that found in animal glycoproteins (Fig. 1). There are, as yet, no reports of plant glycoproteins which contain more than some 40% carbohydrate. In animal glycoproteins, a carbohydrate content of 50% (as found in mucins) to 85% (blood group substances) is not unusual. Most of the sugars commonly found in animal glycoproteins are also common constituents of plant glycoproteins (Table II); exceptions are sialic acid and galac-

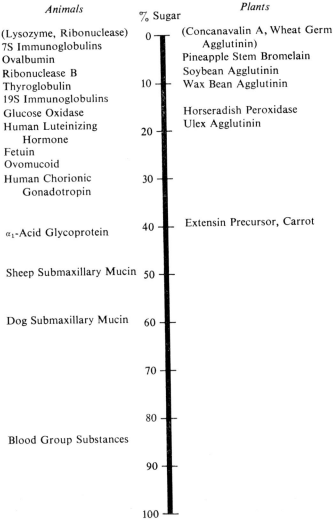

FIG. 1. Spectrum of carbohydrate content of animal and plant glycoproteins.

TABLE II

Sugar constituents of glycoproteins[a]

Sugar	Animals	Plants
Hexoses		
D-Galactose	+	+
D-Glucose	+[b]	+
D-Mannose	+	+
Deoxyhexoses		
L-Fucose	+	+
Hexosamines		
N-Acetyl-D-galactosamine	+	(+)
N-Acetyl-D-glucosamine	+	+
Sialic acids	+	−
Pentoses		
D-Xylose	+[c]	+
L-Arabinose	?	+

[a] In most glycoproteins the configuration of the sugars has not been established.

[b] Limited to the collagens.

[c] In mucopolysaccharides (which include also uronic acids, not listed).

tosamine. Sialic acid has only once been reported to occur in plant seeds (Mayer *et al.*, 1964), but it is not known whether it is linked to protein. Moreover, no confirmation of this report has appeared. Galactosamine has been found only in some of the horseradish peroxidase isozymes (also reported to contain mannosamine which is not present in animal glycoproteins (Shannon *et al.*, 1966)) and in structural glycoproteins of green plastids (Lagoutte and Duranton, 1972). On the other hand, xylose, found so far only in mucopolysaccharides from animals, and arabinose, the occurrence of which in mucopolysaccharides has been questioned (Katzman, 1971), are not uncommon in plant glycoproteins.

With regard to the carbohydrate–protein linkages (Table III), only N-acetyl-D-glucosaminyl-β-L-asparagine (or asparaginyl-N-acetyl-D-glucosamine) has

TABLE III

Carbohydrate–peptide linkages in glycoproteins of animals and plants

Linkage	Animals	Plants
N-Glycosidic		
N-acetyl-D-glucosaminyl-β-L-asparagine	+	+
O-Glycosidic		
N-acetyl-D-galactosaminyl-α-L-serine (or L-threonine)	+	−
D-galactosyl-β-5-hydroxyl-L-lysine	+	−
L-arabinosyl-4-hydroxy-L-proline	−	+

so far been found both in plants and in animals. The other types of linkage appear at present to be confined to one of the kingdoms only. In addition to those listed, an unusual linkage, 4-L-alanyl-D-xylopyranose has been reported to be present in barley albumin (Hochstrasser, 1961) but no confirmation of this report has appeared.

A comprehensive treatise (Gottschalk, 1972) and a number of reviews on glycoproteins (Marshall and Neuberger, 1968, 1970; Marshall, 1972; Montgomery, 1970; Schmid, 1972; Sharon, 1966; Spiro, 1970) have recently appeared. In these, plant glycoproteins are mentioned rather briefly, if at all. Here, I shall review those plant glycoproteins that have been well characterized and for which evidence on the nature of the carbohydrate moieties has been presented. For the sake of convenience, I have classified these glycoproteins according to their biological activities.

III. LECTINS (PHYTOAGGLUTININS)

The largest single group of plant glycoproteins are the lectins, known also as plant agglutinins or phytoagglutinins. These cell-agglutinating and sugar-specific proteins (Sharon and Lis, 1972; Lis and Sharon, 1973) have been known since the turn of the century but have only recently become the focus of intense interest in a large number of laboratories. Lectins exhibit a host of interesting and most unusual chemical and biological properties. Thus, they agglutinate erythrocytes, in certain cases with very high specificity, and some of them are used in the typing of human blood and in the study of the chemical structure of blood group substances. They bind sugars and precipitate polysaccharides and glycoproteins, specifically. Some lectins, such as the one prepared from the red kidney bean (*Phaseolus vulgaris*) and known as PHA, are mitogenic in that they stimulate the conversion of resting lymphocytes into actively growing and dividing blast-like cells. Because of these and other properties, lectins provide a new and most useful tool for the investigation of specific binding sites on protein molecules and serve as models for the study of the antigen–antibody reaction. In addition, they are used for structural studies of carbohydrate-containing polymers and as specific reagents for the isolation of polysaccharides and glycoproteins. Because of their mitogenic activity, they aid in examining the biochemical events involved in the initiation of cell division. Most important, however, is their ability to agglutinate preferentially malignant cells, which has resulted in a growing use of lectins in investigating the architecture of cell surfaces and elucidating the changes which cells undergo upon malignancy, as well as in isolating specific membrane constituents.

Most plant lectins that have been isolated in pure form appear to be glycoproteins (Table IV): this is also true for lectins from other sources (Sharon and Lis, 1972). However, some lectins, such as concanavalin A (Agrawal and Goldstein, 1968; Olson and Liener, 1967), wheat germ

TABLE IV

Glycoprotein lectins from higher plants[a]

Source		Molecular weight	Sub-units	Carbohydrates — Constituents									Mito-genicity	Agglutination of malignant cells	Specificity	
				Neutral sugar[b]	Gal	Glc	Man	Fuc	Ara	Xyl	GlcN	Total %			Human blood type	Sugar
Leguminosae																
Bauhinia purpurea alba					+	+	++	+			++	11	—	—	—	D-GalNAc
Dolichos biflorus (horse gram)		140 000					+				+	3·5	—	—	A	D-GalNAc
Glycine max (soybean)		120 000	4				++			+	+	5·7		+	—	D-GalNAc
Lens culinaris[c] (common lentil)	A	39–48 000	2	++		+					+	2	+	+		α-D-Man
	B	48 000	2	++												
Lotus tetragonolobus	I	120 000					++	++	++		+	9·4			H(O)	α-L-Fuc
	II	58 000					++	++	++		+	4·8			H(O)	α-L-Fuc
	III	117 000					+			+	+	9·2			H(O)	α-L-Fuc
Phaseolus lunatus[d] (lima bean)	I	195–269 000	8								+	4			A	D-GalNAc
	II	110–138 000	4								+[e]	4			A	D-GalNAc
Phaseolus vulgaris (black kidney bean)		128 000					++				++	5·7			—	D-GalNAc
Phaseolus vulgaris (red kidney bean)	E	115–150 000	4 or 8		++	++	+++	+++++	+++	+++	++	8·9	+++	++	—	—
	L	115–140 000	4		++	++	+++	+++++	+++	++++	++	4·1	+++	++	—	—
Phaseolus vulgaris (yellow wax bean)	I	130 000	4				+	+++				10				
	II	130 000	4									10·7				
Robinia pseudoacacia (black locust)		90 000	4													
Ulex europeus (gorse)	I	170 000			++	(+)	++	(+)	+	(+)	+	5·2			H(O)	L-Fuc
	II						++				+	21·7			H(O)	(D-GlcNAc)$_2$
Other plants																
Phytolacca americana (pokeweed)		32 000			++	+	++	+	+		++	~5	+		—	—
Solanum tuberosum (potato)												5·2				(D-GlcNAc)$_2$
Wistaria floribunda		($S_w = 4·5$)					++	+	+	+	++	11·4	+			D-GalNAc

[a] Data for Bauhinia purpurea alba from Irimura and Osawa (1972); for yellow wax bean from Takahashi et al. (1967, 1968) and Sela et al. (1973); for Wistaria floribunda mitogen from Toyoshima et al. (1971); for pokeweed from Reisfeld et al. (1967) and Börjeson et al. (1966). For source of other data, see Table 4 in Sharon and Lis, 1972.
[b] Sugar not identified.
[c] Also known as Lens esculenta.
[d] Also known as Phaseolus limensis.
[e] Hexosamine, not identified.

agglutinin (Allen *et al.*, 1973) and possibly also the agglutinin of the garden pea (Entlicher *et al.*, 1970), do not contain covalently bound sugar and thus are not glycoproteins. It would therefore appear that the sugar moiety of lectins plays no role in their biological activity.

The role of lectins in nature, whether in plants or in other organisms, is still a mystery (Sharon and Lis, 1972). It has been suggested that they are antibodies intended to counteract soil bacteria; that they serve to protect plants against fungal attack by inhibiting fungal polysaccharidases; that because of their affinity for saccharides they are involved in sugar transport and storage; or that they serve for the attachment of glycoprotein enzymes in organized multienzyme systems. In view of the mitogenic properties of lectins it is possible that their function is to control cell division and germination in plants. There is, however, no evidence for or against any of these hypotheses. Indeed, it has been proposed that the biological properties of lectins, as observed in the laboratory, have no relation to their function in nature.

Of the many glycoprotein lectins only two have been investigated in any detail as to the chemistry of their carbohydrate moieties. These are soybean agglutinin and wax bean agglutinin.

A. SOYBEAN AGGLUTININ (SBA)

Soybean agglutinin was first isolated in purified form and studied by Liener and his coworkers (Liener and Pallansch, 1952; Wada *et al.*, 1958). Lis *et al.* (1966b) haved eveloped improved methods for isolating highly purified SBA, first by the use of chromatography on hydroxylapatite and very recently by affinity chromatography on a column of ϵ-aminocaproyl-β-D-galactopyranosylamine coupled to Sepharose (Gordon *et al.*, 1972). The molecular weight of SBA was 110 000 by ultracentrifugation (Wada *et al.*, 1958; Lis *et al.*, 1966b) and 120 000 by gel filtration (Lotan *et al.*, 1974). Electrophoresis on polyacrylamide gel and gel filtration, both in the presence of SDS, showed that SBA is a tetramer with four apparently identical subunits, each of molecular weight 30 000 $\pm$ 500 (Lotan *et al.*, 1974).

SBA was one of the first plant glycoproteins to be recognized and characterized. It contains 4·5% mannose and 1·2% *N*-acetyl-D-glucosamine (Lis *et al.*, 1966b; H. Lis and N. Sharon, unpublished). After digestion of SBA by pronase, the carbohydrate moiety was isolated in the form of a glycopeptide containing only aspartic acid, D-mannose, and *N*-acetyl-D-glucosamine, in the ratio of 1:9:2–3 (Lis *et al.*, 1966b; H. Lis and N. Sharon, unpublished). Estimation of the molecular weight of the glycopeptide by ultracentrifugation, gel filtration, and end group analysis gave a value of approximately 2000 (H. Lis and N. Sharon, unpublished), not 4500 as previously reported (Lis *et al.*, 1966b). Thus the lectin probably contains four carbohydrate chains, one per subunit.

The carbohydrate–peptide linking group has been isolated and shown to be identical in its composition and behaviour on paper chromatography and electrophoresis with *N*-acetyl-D-glucosaminyl-β-L-asparagine (Lis *et al.*, 1969). Studies still in progress (H. Lis and N. Sharon, unpublished) have led to the assignment of the following tentative structure for the glycopeptide:

$$(\text{Man})_{4-5}\text{GlcNAc}(\text{Man})_{5-4}(\text{GlcNAc})_{1-2}\text{Asn}$$

Most, or all, of the D-mannose residues are α-linked and the *N*-acetyl-D-glucosamine residues are β-linked. Of particular interest is the general similarity of this glycopeptide to glycopeptides isolated from certain animal glycoproteins, such as ovalbumin and ribonuclease B (Marshall, 1972; Spiro, 1970; see Fig. 2).

In addition to the soybean agglutinin described above, soybean oil meal was found to contain three minor hemagglutinins (Lis *et al.*, 1966a). The four agglutinins or isolectins (Sharon and Lis, 1972) are separable on columns of DEAE-cellulose but behave identically upon chromatography on hydroxyl-apatite and carboxymethylcellulose, as well as in acrylamide gel electrophoresis. They all contain mannose and glucosamine (Lis *et al.*, 1966a). The occurrence of isolectins has been observed in many other seeds (Sharon and Lis, 1972), such as lentil (*Lens culinaris*) (Howard *et al.*, 1971; Ticha *et al.*, 1970), lima bean (*Phaseolus limensis*) (Galbraith and Goldstein, 1972; Gould and Scheinberg, 1970) and wheat (*Triticum vulgare*) (Allen *et al.*, 1973). Such isolectins may be the product of closely related genes, or they may be formed prior to or during isolation as a result of side chain modifications, such as hydrolysis of the amide groups of glutamine or asparagine in the protein. In the isolectins that are glycoproteins, the differences may reside in the carbohydrate side chains.

Agglutination of erythrocytes or malignant tissue culture cells by SBA is strongly and specifically inhibited by low concentrations of *N*-acetyl-D-galactosamine ($0.05–0.1$ μmol/ml), by disaccharides with *N*-acetyl-D-galactosamine at their nonreducing ends or to a lesser degree by D-galactose (Lis *et al.*, 1970). Equilibrium dialysis and gel filtration on columns of Sephadex in equilibrium with the radioactive ligand both showed that SBA contains two identical binding sites for *N*-acetyl-D-galactosamine per 120 000 daltons, with a K_a value of 3.0×10^4 1/mol (Lotan *et al.*, 1973).

B. WAX BEAN AGGLUTININ (WBA)

The lectin of the yellow wax bean (*Phaseolus vulgaris*) was originally purified by Takahashi *et al.* (1967) using fractionation with ammonium sulfate and successive chromatography on DEAE- and carboxymethyl-cellulose. The purified material was homogeneous by ultracentrifugal analysis and by electrophoresis on polyacrylamide gels. It has a molecular weight of 132 000 and contains 10.4% covalently bound carbohydrate, of which mannose and glucose are the predominant sugars, with lesser amounts of glucosamine, galactose, arabinose, xylose, and fucose. A glycopeptide, isolated from a pronase digest of WBA (Takahashi and Liener, 1968) had a molecular weight of about 4380; it contained 12 amino acid residues (6 of which were aspartic acid residues) and 19 sugar residues with the composition $Man_5Ara_5Gal_4$-$GlcN_2Glc_2Fuc_1$.

Since the molar ratios of the sugars in the glycopeptide are quite different from those found in the intact molecule of WBA ($Man_{38}Ara_9Gal_4GlcN_6$-$Glc_{14}Fuc_3Xyl_3$), it seems reasonable to assume that WBA contains two or more complex heterosaccharide units of different composition and structure (Takahashi and Liener, 1968). Although the aspartic acid and threonine residues in the glycopeptide isolated from WBA would provide suitable loci for the covalent attachment of carbohydrate units, the precise nature of the carbohydrate–peptide linkage in WBA remains to be elucidated.

Recently, the material obtained according to Takahashi et al. (1967) was further fractionated (Sela et al., 1973) by chromatography on hydroxylapatite. Three fractions were obtained, two of which were biologically active. They were hemagglutinating and mitogenic and caused detectable agglutination of transformed mammalian cells at extremely low concentrations (0·2–0·5 μg/ml). The concentrations required for the agglutination of the normal parent cells were about 100 times higher. One of the two fractions was also tested on animals and found to inhibit tumor development. The two fractions showed no significant differences in amino acid and sugar composition, nor in their molecular weights (130 000 ± 5000, as estimated by gel filtration). Polyacrylamide gel electrophoresis in the presence of sodium dodecyl sulphate suggested that both fractions are made up of a single type of subunit of 30 000 molecular weight.

From early inhibition studies carried out with crude preparations of WBA and crude neuraminidase it was concluded (Liener and Northrop, 1959) that the lectin binds to sialic acid-like sites on the surface of erythrocytes. Fetuin was an extremely potent inhibitor of the agglutination reaction (Sela et al., 1973) and the inhibitory action was markedly increased when fetuin was heated for 3 min at 100°C; removal of sialic acid, on the other hand, did not affect the inhibitory power of fetuin. It seems, therefore, that the inhibitory activity of fetuin does not result from interaction of WBA with its terminal sialic acid but rather with some inner structure of the carbohydrate chain. Sialic acid is thus most probably not involved in the binding of WBA to cells.

IV. ENZYMES

Only one plant enzyme, pineapple stem bromelain, has been conclusively shown to be a glycoprotein, by the isolation of a glycopeptide from its proteolytic digests. For other plant glycoprotein enzymes such evidence is still not available, but since they have been purified to a high degree the presence of sugars strongly suggests that these enzymes are glycoproteins. The latter include ascorbic acid oxidase, which contained 1·4% glucosamine (Stark and Dawson, 1962), and horseradish (*Armaracta rusticana*) peroxidases (Shannon *et al.*, 1966). Seven such peroxidases were isolated, all of which contain 15–18% sugars. In addition to sugars commonly found in plant glycoproteins (galactose, fucose, mannose, arabinose and xylose) some of the peroxidase isozymes were reported to contain galactosamine and mannosamine. The latter has not been encountered in any other glycoprotein, and confirmation of this report will be awaited with interest.

A. PINEAPPLE STEM BROMELAIN

Proteolytic enzymes, known as bromelains, have been isolated from the stem and fruit of the pineapple. They are thiol proteases, and in this respect resemble papain and ficin, which are also found in plants. However, papain and ficin are not glycoproteins.* From crude stem bromelain several fractions have been obtained which exhibited proteolytic activity and which contained hexose and hexosamine (Murachi *et al.*, 1964; Ota *et al.*, 1964; Feinstein and Whitaker, 1964). Two of the most abundant, enzymatically active components were purified to electrophoretic homogeneity (Scocca and Lee, 1969). They contained the same carbohydrate components, D-glucosamine, D-mannose, D-xylose and L-fucose, in the molar ratios of 2:2:1:1. By exhaustive proteolysis of the bromelains, glycopeptides containing only aspartic acid, glutamic acid and serine, in addition to the sugars, were obtained. Periodate oxidation, methylation and digestion by glycosidases showed that all the carbohydrate of bromelain is in the form of a single branched oligosaccharide chain in which all of the neutral sugars are in nonreducing terminal positions. The two *N*-acetyl-D-glucosamine residues occur in internal positions, in the form of a β-(1 → 4)-linked disaccharide, which is attached through its reducing group to the amide group of asparagine. All the D-mannose residues except one are α-linked; the β-linked residue is attached to the 4 position of the *N*-acetylglucosamine at the nonreducing end of the internal GlcNAc-β(1 → 4)-GlcNAc disaccharide. Such a core structure (Fig. 2) is also present in glycopeptides from ovalbumin, from *Aspergillus oryzae* α-amylase and from ribonuclease B and has been suggested to be of general occurrence (Lee and Scocca, 1972).

In another study of the stem bromelain (Murachi *et al.*, 1967; Yasuda *et al.*, 1970), the carbohydrate moiety was isolated from pronase digests of the parent protein. It was composed of *N*-acetyl-D-glucosamine, D-mannose, D-xylose and L-fucose in the molar ratio of 2:2–3:1:1. From detailed chemical and enzymic studies a structure was proposed for the carbohydrate moiety of bromelain (Fig. 2).

*Ficin has very recently been shown to be a glycoprotein (B. Friedenson and I. E. Liener, 1974, *Biochim biophys. Acta* **342**, 209).

1. Soybean Agglutinin (Lis and Sharon, unpublished)
   $[\alpha\text{-D-Man}]_{5-4}\beta\text{-D-GlcNAc}[\alpha\text{-D-Man}]_{4-5}\beta\text{-D-GlcNAc}_{1-2}\text{-Asn}$

2. Bromelain (Yasuda *et al.*, 1970)

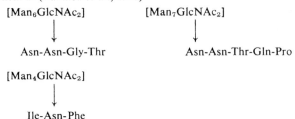

3. Bromelain, Core Structure (Lee and Scocca, 1972)
   D-Man β-(1 → 4)-D-GlcNAc β-(1 → 4)-D-GlcNAc β-(1 →) Asn

4. Ricin D (Funatsu *et al.*, 1971)

   [Man₆GlcNAc₂]   [Man₇GlcNAc₂]

   Asn-Asn-Gly-Thr   Asn-Asn-Thr-Gln-Pro

   [Man₄GlcNAc₂]

   Ile-Asn-Phe

FIG. 2. Tentative structures of glycopeptides from plant glycoproteins.

It will be noted that both the composition and the structure of this carbohydrate moiety differ from those described earlier. In view of the suggestion
that the pineapple produces a family of proteolytic enzymes it is possible that
the two groups of workers studied different pineapple bromelains. However,
in evaluating these results, the great difficulties involved in the structural
elucidation of carbohydrate side chains of glycoproteins should be borne in
mind. These originate mainly from the microheterogeneity of the carbohydrate moieties, from the inability to subfractionate certain heterogeneous
glycopeptide preparations, and from the lack of sensitive criteria for assessing
the degree of homogeneity of such preparations (Marshall and Neuberger,
1970; Schmid, 1972).

Periodate oxidation studies of bromelain seem to indicate that the neutral
sugars of the carbohydrate moiety are not essential for the catalytic activity
of the enzyme (Yasuda *et al.*, 1971).

V. TOXINS AND ALLERGENS

Both the agglutinating and toxic proteins purified from the seeds of castor
bean (*Ricinus communis*) were reported to be glycoproteins, each containing

3·7% neutral sugar, mainly as mannose and galactose (Waldschmidt-Leitz and Keller, 1970). In another study (Funatsu et al., 1971), the toxin ricin D was found to contain mannose as the only neutral sugar, in addition to glucosamine. The reason for the differences in the sugar composition reported by the two groups for the castor bean toxin is not clear. A tentative partial structure of glycopeptides isolated from ricin D has been presented (Fig. 2) (Funatsu et al., 1971).

Two major ryegrass pollen allergens contain mannose, galactose and xylose (Johnson and Marsh, 1966). A glycoprotein allergen was also isolated from short ragweed pollen (Underdown and Goodfriend, 1969), which contained 4·9% hexose, 8·7% pentose and 0·8% hexosamine.

VI. STRUCTURAL PROTEINS—EXTENSIN

Since the original discovery that extensin, a protein rich in the unusual amino acid *trans*-4-hydroxy-L-proline, is a stable, bound component of sycamore callus cell walls (Lamport and Northcote, 1960), speculation has abounded regarding its nature, mode of biosynthesis and role in the cell. However, investigation of proteins of cell walls and cell membranes of plants is made difficult by problems of solubility, since most of them seem to be cross-linked to cell wall polysaccharides. In the course of studies on primary plant cell walls, prepared from cell-suspension cultures, Lamport (1967, 1969) obtained a number of hydroxyproline glycosides by partial alkaline degradation of dry walls from cell suspensions of tomato. The hydroxyproline glycosides, separated by successive fractionation on Sephadex G-25 and ion exchange chromatography, were identified as tri- and tetra-saccharides of arabinose that were each attached by glycosidic linkages to a single hydroxyproline residue. The linkage was assumed to be an *O*-glycosidic bond from the reducing group of the arabinose oligosaccharide, to the 4-hydroxy group of hydroxyproline, as indicated by its alkaline stability and the existence of a free carboxyl and a free secondary amino group of the hydroxyproline residue. Since hydroxyproline accounts for about 30% of all the amino acid residues in tomato cell walls and 70% of the hydroxyproline was released in the form of glycosidic derivatives, it was concluded that most, if not all, of the numerous hydroxyproline residues are involved in this carbohydrate-peptide linkage. Heath and Northcote (1971) further characterized this glycoprotein in sycamore and found that a glycopeptide that had been extracted by hydrazinolysis contained only hydroxyproline, arabinose and galactose.

The suggestion that the hydroxyproline-rich glycoprotein plays a fundamental role in cell wall structure receives support from recent work which shows the occurrence of hydroxyproline arabinosides in cell walls of many plants, ranging from the spermatophytes to the green alga *Chlorella* (Lamport and Miller, 1971). Hydroxyproline arabinosides have also been isolated from one of the horseradish peroxidase isozymes (Liu and Lamport, 1968).

To date, no other hydroxyproline glycosides have been isolated from higher plants. It is, however, interesting to note that hydroxyproline-O-galactoside has been found in hydrolysates of a crude cell wall fraction of the alga *Chlamydomonas reinhardtii* (Miller *et al.*, 1972).

An intact precursor of extensin has been isolated from carrot disks: before this material is actually incorporated into the cell wall matrix it is transiently associated with membranous organelles of the cytoplasm and can be extracted from the wall with salt solutions, hence isolation is not difficult (Brysk and Crispeels, 1972). The precursor accounts for one third of the total hydroxyproline secreted into the wall matrix. It had a sedimentation constant of 4·5S and appeared to be composed of 60% protein and 40% carbohydrate. The carbohydrate component was primarily arabinose, and the principal amino acids were serine, hydroxyproline, lysine and glycine.

In addition to the specific cell wall protein, extensin, there is a number of other and different classes of plant glycoproteins containing hydroxyproline. Several phenol-soluble glycoproteins, each with characteristic and different composition, molecular size and solubility, have been isolated by chromatographic procedures from the leaves of *Vicia faba*. The major glycoproteins contained 9–14% galactose and arabinose as the main sugar components and small, variable amounts of glucose, mannose, xylose, fucose and rhammose. Hydroxyamino acids (hydroxyproline, threonine, serine) were the predominant amino acid components of the protein moiety (Pusztai and Watt, 1969).

In the case of a protein–polysaccharide complex extracted from corn pericarp (Boundy *et al.*, 1967) about 10% of the total amino acid residues consisted of hydroxyproline. The polysaccharide portion of this material contained glucose in major amounts, along with glucosamine, galactose, arabinose and xylose.

VII. OTHER PLANT GLYCOPROTEINS

A. 7S PROTEIN OF SOYBEAN

In addition to soybean agglutinin, described above, another glycoprotein (the 7S protein) is also present in soybeans (Koshiyama, 1965). The latter, which comprises about 20% of the total protein of the soybean (Wolf, 1970), was purified from the soybean globulin fraction, and obtained in a form that was homogeneous when examined by electrophoresis, chromatography on hydroxylapatite, gel filtration on Sephadex G-200 and in the ultracentrifuge (Koshiyama, 1965, 1966, 1968a,b, 1972). Its molecular weight was found by four different methods to be 180–210 000 (Koshiyama, 1968b); it undergoes reversible dimerization to a 9S protein with a molecular weight of about 370 000 (Koshiyama, 1968b,c,d), and it is made up of subunits of molecular weight 22–24 000 (Koshiyama, 1971). The 7S protein had nine N-terminal residues and contained 3·8% mannose (38 residues/mole) and 1·2% glucosamine (12 residues/mole) (Koshiyama, 1967, 1968b). No sialic or uronic acid was detected.

Further evidence for the glycoprotein nature of the 7S protein was obtained by the isolation of a glycopeptide from its pronase digest (Koshiyama, 1969). The glycopeptide, which was homogeneous by amino-terminal analysis and high voltage electrophoresis, had a molecular weight of about 9870 and was composed of 17 amino acid residues ($Asp_6Thr_3Ala_3Gly_3Glu_1Ser_1$) and 51 sugar residues ($Man_{39}GlcN_{12}$). The nature of the carbohydrate–peptide linkage was not established.

Since only one kind of glycopeptide was obtained, the carbohydrate content of which corresponded to that of the original protein, it was concluded (Koshiyama, 1969) that the carbohydrate moiety is present in the protein as a single polysaccharide unit. If so, this carbohydrate moiety (molecular weight 8000–8500) is considerably larger than any reported to date in the literature; the carbohydrate moieties of typical glycoproteins described in the past contain at most 15 saccharides per chain (Marshall and Neuberger, 1970; Marshall, 1972). The result of further studies on this glycopeptide from the 7S soybean protein will be awaited with interest.*

B. KIDNEY BEAN GLYCOPROTEINS

Two glycoproteins have been isolated in a highly purified form from the seeds of kidney bean (*Phaseolus vulgaris*). Glycoprotein I (Pusztai, 1965, 1966), which represents nearly 0·5% of total seed nitrogen, contained 11·8% neutral sugar, with mannose (7·8%) being the predominant component, together with small amounts of pentoses and methylpentoses. In addition, glycoprotein I contained 2·7% glucosamine. Glycoprotein II (Pusztai and Watt, 1970) is one of the largest single glycoprotein constituents from the seeds of kidney bean, accounting for at least 10% of the total nitrogen of the seed. This glycoprotein occurs as a monomer with molecular weight 140 000 and, between pH 3·4 and 6·6, as a tetramer, molecular weight 560 000. The carbohydrate part is composed mainly of mannose (3·2%) and glucosamine (1%). In contrast to a number of glycoproteins isolated from different cultivars of *Phaseolus vulgaris* (Sharon and Lis, 1972; Lis and Sharon, 1973), this glycoprotein had no hemagglutinating activity for rabbit erythrocytes.

VIII. CONCLUDING REMARKS

The study of plant glycoproteins is still in its infancy. Although some structural features of these compounds are beginning to emerge, many more

---

* Three kinds of glycopeptides with a composition of $Asp_1GlcN_2Man_{6-8}$ have very recently been isolated from the 7S protein of soybean by F. Yamauchi following more complete digestion of the protein with pronase; the carbohydrate of this glycoprotein moiety appears, therefore, to be made up of a number of comparatively small side chains, which are linked to the protein in several positions along the polypeptide chain (I. Koshiyama, private communication, 1973).

plant glycoproteins need to be isolated and characterized, before any firm generalizations can be made as to their chemical constitution. Virtually no studies have been carried out on the biosynthesis of plant glycoproteins. It has been shown that glucosamine can be incorporated by plant tissues into the amino sugar moieties of glycoprotein-like substances (Roberts, 1970; Roberts *et al.*, 1971, 1972), and it may be safely assumed that glycoprotein biosynthesis in plants does not differ markedly from that of their animal counterparts. Last, but not least, we know next to nothing about the function in nature of most plant glycoproteins, be they lectins, toxins or other proteins, and no clear role for their carbohydrate side chains has so far emerged.

It is hoped that the renewed interest in plant biochemistry will lead, among other things, to systematic studies of plant glycoproteins. Such studies will, undoubtedly, provide much needed insight into the chemistry and physiological functions of these compounds and may also contribute to the improved utilization of plants by man.

ACKNOWLEDGEMENT

I wish to thank my colleague, Dr Halina Lis, for her help in the preparation of this manuscript.

REFERENCES

Agrawal, B. B. L. and Goldstein, I. J. (1968). *Archs Biochem. Biophys.* **124**, 218.

Allen, A. K., Neuberger, A. and Sharon, N. (1973). *Biochem. J.* **131**, 155.

Baldwin, E. (1949). "Comparative Biochemistry", Cambridge University Press, London and New York.

Börjeson, J., Reisfeld, R. A., Chessin, L. N., Welsh, P. D. and Douglas, S. D. (1966). *J. exp. Med.* **124**, 859.

Boundy, J. A., Wall, J. S., Turner, J. E., Woychik, J. H. and Dimler, R. J. (1967). *J. biol. Chem.* **242**, 2410.

Brysk, M. M. and Crispeels, M. J. (1972). *Biochim. biophys. Acta* **257**, 421.

Entlicher, G., Koštíř, J. V. and Kocourek, J. (1970). *Biochim. biophys. Acta* **221**, 272.

Feinstein, G. and Whitaker, J. R. (1964). *Biochemistry* **3**, 1050.

Funatsu, M., Ishiguro, M., Nanno, S. and Hara, K. (1971). *Proc. Jap. Acad.* **47**, 718.

Galbraith, W. and Goldstein, I. J. (1972). *Biochemistry* **11**, 3976.

Gordon, J. A., Blumberg, S., Lis, H. and Sharon, N. (1972). *FEBS Letters* **24**, 193.

Gottschalk, A. (1972). Ed. "Glycoproteins", BBA Library, Vol. 5, 2nd ed. Elsevier, Amsterdam.

Gould, N. R. and Scheinberg, S. L. (1970). *Archs. biochem. Biophys.* **137**, 1.

Heath, M. F. and Northcote, D. H. (1971). *Biochem. J.* **125**, 953.

Hochstrasser, K. (1961). *Z. physiol. Chem.* **324**, 250.

Howard, I. K., Sage, H. J., Stein, M. D., Young, N. M., Leon, M. A. and Dyckes, D. F. (1971). *J. biol. Chem.* **246**, 1590.

Irimura, T. and Osawa, T. (1972). *Archs Biochem. Biophys.* **151**, 475.
Johnson, P. and Marsh, D. G. (1966). *Immunochemistry* **3**, 101.
Katzman, R. L. (1971). *J. Neurochem.* **18**, 1187.
Koshiyama, I. (1965). *Agric. biol. Chem.* **29**, 885.
Koshiyama, I. (1966). *Agric. biol. Chem.* **30**, 646.
Koshiyama, I. (1967). *Agric. biol. Chem.* **31**, 874.
Koshiyama, I. (1968a). *Agric. biol. Chem.* **33**, 281.
Koshiyama, I. (1968b). *Cereal Chem.* **45**, 394.
Koshiyama, I. 1968c). *Agric. biol. Chem.* **32**, 879.
Koshiyama, I. (1968d). *Cereal Chem.* **45**, 405.
Koshiyama, I. (1969). *Archs Biochem. Biophys.* **130**, 370.
Koshiyama, I. (1971). *Agric. biol. Chem.* **35**, 385.
Koshiyama, I. (1972). *Agric. biol. Chem.* **36**, 2255.
Lagoutte, B. and Duranton, J. (1972). *FEBS Letters* **28**, 333.
Lamport, D. T. A. (1967). *Nature, Lond.* **216**, 1322.
Lamport, D. T. A. (1969). *Biochemistry* **8**, 1155.
Lamport, D. T. A. and Miller, D. H. (1971). *Pl. Physiol.* **48**, 454.
Lamport, D. T. A. and Northcote, D. H. (1960). *Nature, Lond.* **188**, 665.
Lee, Y. C. and Scocca, J. R. (1972). *J. biol. Chem.* **247**, 5753.
Liener, I. E. and Northrop, R. L. (1959). *Proc. Soc. exp. Biol. Med.* **100**, 105.
Liener, I. E. and Pallansch, M. J. (1952). *J. biol. Chem.* **197**, 29.
Lis, H., Fridman, C., Sharon, N. and Katchalski, E. (1966a). *Archs Biochem. Biophys.* **117**, 301.
Lis, H. and Sharon, N. (1973). *A. Rev. Biochem.* **42**, 541.
Lis, H., Sharon, N. and Katchalski, E. (1966b). *J. biol. Chem.* **241**, 684.
Lis, H., Sharon, N. and Katchalski, E. (1969). *Biochim. biophys. Acta* **192**, 364.
Lis, H., Sela, B. A., Sachs, L. and Sharon, N. (1970). *Biochim. biophys. Acta* **211**, 582.
Liu, E. and Lamport, D. T. A. (1968). *Pl. Physiol.* **43**, S-16.
Lotan, R., Siegelman, H. W., Lis, H. and Sharon, N. (1974). *J. biol. chem.* **249**, 1219.
Marshall, R. D. (1972). *A. Rev. Biochem.* **41**, 673.
Marshall, R. D. and Neuberger, A. (1968). *In* "Carbohydrate Metabolism and its Disorders" (F. Dickens, P. J. Randle and W. J. Whelan, eds), Vol. I, p. 213. Academic Press, New York and London.
Marshall, R. D. and Neuberger, A. (1970). *Adv. Carbohydr. Chem.* **25**, 407.
Mayer, F. C., Dam, R. and Pazur, J. H. (1964). *Archs Biochem. Biophys.* **108**, 356.
Miller, D. H., Lamport, D. T. A. and Miller, M. (1972). *Science, N.Y.* **176**, 918.
Montgomery, R. (1970). *In* "The Carbohydrates" (W. Pigman and D. Horton, eds), Vol. IIB, p. 627. Academic Press, New York and London.
Murachi, T., Yasui, M. and Yasuda, Y. (1964). *Biochemistry* **3**, 48.
Murachi, T., Suzuki, A. and Takahashi, N. (1967). *Biochemistry* **6**, 3730.
Neuberger, A. (1971). *In* "Glycoproteins of Blood Cells and Plasma" (G. A. Jamieson and T. J. Greenwalt, eds), p. 1. J. B. Lippincott, Philadelphia and Toronto.
Olson, M. O. J. and Liener, I. E. (1967). *Biochemistry* **6**, 105.
Ota, S., Moore, S. and Stein, W. H. (1964). *Biochemistry* **3**, 180.
Pusztai, A. (1965). *Biochem. J.* **95**, 3c.
Pusztai, A. (1966). *Biochem. J.* **101**, 379.
Pusztai, A. and Watt, W. B. (1969). *Eur. J. Biochem.* **10**, 523.

Pusztai, A. and Watt, W. B. (1970). *Biochim. biophys. Acta* **207**, 413.

Reisfeld, R. A., Börjeson, J., Chessin, L. N. and Small, P. A., Jr. (1967). *Proc. natn. Acad. Sci. U.S.A.* **58**, 2020.

Roberts, R. M. (1970). *Pl. Physiol.* **45**, 263.

Roberts, R. M., Connor, A. B. and Cetorelli, J. J. (1971). *Biochem. J.* **125**, 999.

Roberts, R. M., Cetorelli, J. J., Kirby, E. G. and Ericson, M. (1972). *Pl. Physiol.* **50**, 531.

Schmid, K. (1972). *Chimia* **26**, 405.

Scocca, J. R. and Lee, Y. C. (1969). *J. biol. Chem.* **244**, 4852.

Sela, B. A., Lis, H., Sharon, N. and Sachs, L. (1973). *Biochim. biophys. Acta* **310**, 273.

Shannon, L. M., Kay, E. and Lew, J. Y. (1966). *J. biol. Chem.* **241**, 2166.

Sharon, N. (1965). *In* "The Amino Sugars" (R. W. Jeanloz and E. A. Balazs, eds), Vol. IIA, p. 1. Academic Press, New York and London.

Sharon, N. (1966). *A. Rev. Biochem.* **35**, 485.

Sharon, N. and Lis, H. (1972). *Science, N.Y.* **177**, 949.

Spiro, R. G. (1970). *A. Rev. Biochem.* **39**, 599.

Stark, G. R. and Dawson, C. R. (1962). *J. biol. Chem.* **237**, 712.

Takahashi, T. and Liener, I. E. (1968). *Biochim. biophys. Acta* **154**, 560.

Takahashi, T., Ramachandramurthy, P. and Liener, I. E. (1967). *Biochim. biophys. Acta* **133**, 123.

Tichá, M., Entlicher, G., Koštíř, J. V. and Kocourek, J. (1970). *Biochim. biophys. Acta* **221**, 282.

Toyoshima, S., Akiyama, Y., Nakano, K., Tonomura, A. and Osawa, T. (1971). *Biochemistry* **10**, 4457.

Underdown, B. J. and Goodfriend, L. (1969). *Biochemistry* **8**, 980.

Wada, S., Pallansch, M. J. and Liener, I. E. (1958). *J. biol. Chem.* **233**, 395.

Waldschmidt-Leitz, E. and Keller, L. (1970). *Z. Physiol. Chem.* **351**, 490.

Wolf, W. J. (1970). *J. agric. Fd. Chem.* **18**, 969.

Yasuda, Y., Takahashi, N. and Murachi, T. (1970). *Biochemistry* **9**, 25.

Yasuda, Y., Takahashi, N. and Murachi, T. (1971). *Biochemistry* **10**, 2624.

Author Index

Numbers in italics are those pages on which references are listed

A

Adams, G. A., 154, *163*
Agrawal, B. B. L., 240, *250*
Akazawa, T., 11, *26*, 135, 141, 142, *143*, *144*
Akiyama, Y., 241, *252*
Alam, S. S., 203, *203*
Albersheim, P., 64, 73, 75, 76, *80*, *81*, 146, 147, 148, 149, 150, 151, 152, 153, 154, 155, 156, 157, 158, 159, 160, 161, *163*, *164*, 183, *189*, 194, 203, *204*, *205*
Albrecht, G. J., 141, *144*
Alexander, M., 220, *232*
Allan, B. J., 114, 115, *124*
Allcock, C., 102, *107*
Allen, A. K., 242, 243, *250*
Anagnostopoulos, C., 3, *5*
Anderson, B., 223, *231*
Anderson, J. D., 50, 56, *59*
Andrews, P., 50, 56, *59*
Andrews, T. J., 21, *25*, *26*, 44, *45*
Applegarth, D. A., 220, *231*
Arnold, R., 196, 197, *204*
Aronson, J. M., 220, *231*
Asai, T., 49, *59*
Asami, K., 162, *164*
Aspinall, G. O., 154, 158, *163*
Avai, Y., 228, *232*
Avigad, G., 193, *204*
Avron, M., 162, *164*
Axelrod, B., 31, *45*, 56, *59*
Axelsson, K., 220, *231*

B

Baardseth, E., 213, *217*
Bäckstrom, G., 215, *217*
Bacon, J. S. D., 220, 228, *231*, *232*
Badenhuizen, N. P., 117, *124*, 129, 134, *143*, *144*
Bailey, R. W., 195, *203*

Baker, E. A., 95, *96*
Baldry, C. W., 20, *25*, 68, 70, *80*
Baldwin, E., 235, *250*
Ballou, C. E., 226, *232*
Bamberger, E. S., 95, *95*
Banks, B. E. C., 9, *25*
Banks, W., 110, 115, 121, *124*
Barber, G. A., 186, *189*, 209, *216*, *217*, *218*
Barham, D., 85, *95*
Barker, S. A., 227, 229, *231*
Barry, V. C., 227, *232*
Bartnicki-Garcia, S., 219, 221, 222, 227, 229, 231, *232*, 233
Bass, S. T., 141, *144*
Bassham, J. A., 7, 9, 11, *25*, *26*, 139, *143*
Bateman, D. F., 147, *163*
Bathgate, G. N., 122, 123, *124*, *125*
Batt, R. D., 102, *107*
Bau, A., 84, *95*
Bauer, W. D., 146, 147, 148, 149, 150, 151, 152, 153, 154, 155, 156, 157, 158, 160, 161, *163*, *164*, 183, *189*
Baun, L. C., 130, 134, *143*
Becker, G. E., 146, *163*
Becker, W. M., 29, *46*
Bednar, T. W., 50, *59*
Beevers, H., 27, 28, 29, 30, 31, 35, 36, 40, 45, *45*, *46*
Begbie, R., 154, *163*
Behrens, N. H., 199, 201, *204*
Bell, D., 227, *232*
Benson, A. A., 99, 105, 106, *107*, *108*
Berger, L. R., 220, *232*
Bettiger, H., 173, *180*
Biale, J. B., 103, 104, *107*
Bidwell, R. G. S., 50, 51, *59*
Bilderback, D. E., 115, *124*
Bird, I. F., 72, 79, *80*, 135, *143*
Bishop, C. T., 154, *163*
Björndal, H., 146, *163*, 220, *231*

Nojima, S., 97, *107*
Nordin, J. H., 230, *232*
Northcote, D. H., 40, *46*, 166, 167, 168,
169, 170, 172, 173, 174, 175, 176, 177,
178, *179*, *180*, *181*, 192, *204*, 222, 223,
227, *232*, *233*, 247, *250*, *251*
Northrop, R. L., 244, *251*
Novellie, L., 113, 114, 115, *124*
Nuccorini, R., 49, *59*

O

Obaidah, M. A., 227, 230, *232*
Ochoa, S., 10, *26*
Odzuck, W., 193, 196, 197, 198, *204*
Ogren, W. L., 21, *25*
Ohira, K., 194, *204*
Ohno, K., 97, *107*
Ohsaka, Y., 228, *232*
Olson, M. O. J., 240, *251*
O'Mant, D. M., 229, *231*
Ongun, A., 104, 105, *107*
Ordin, L., 186, *189*
Osawa, T., 241, *251*, *252*
Osmond, C. B., 22, *26*, 135, *144*
Osterhout, W. J. V., 20, *26*
Ota, S., 245, *251*
Ozbun, J. L., 130, 141, *144*

P

Painter, T. J., 216, *217*, 231, *232*
Paju, J., 48, *59*
Pakhomova, J. V., 208, *217*
Palade, G. E., 167, 179, *179*, *180*, *181*
Palan, P. R., 90, 92, 94, *95*
Pallansch, M. J., 242, *251*, *252*
Palmer, G. H., 123, *124*, *125*
Palmiano, E. P., 130, 134, *143*
Panayotatos, N., 184, *189*
Park, R. B., 95, *95*, 105, *107*
Parodi, A. J., 199, 201, *204*
Patterson, A. A., 24, *26*
Patterson, J. C., 227, *232*
Payne, D. M., 110, *125*
Pazur, J. H., 239, *251*
Peat, S., 119, *124*, 227, *232*
Pecker, M. M., 106, *107*
Percheron, F., 4, 5, *5*, *6*
Percival, E., 208, 213, *217*
Perez, C. M., 130, 134, *143*
Petek, F., 3, *5*, 84, 85, *95*, *96*

Pickett-Heaps, J. D., 168, 169, 176, 178,
181
Plouvier, V., 48, *59*
Pohl, P., 101, *108*
Poincelot, R. P., 16, *26*
Poole, A. G., 101, *107*
Porter, D. W., 110, *125*
Porter, H. K., 72, 79, *80*, 129, 139, 140,
141, *144*
Potter, P. K., 129, *144*
Preiss, J., 65, 78, *80*, *81*, 130, 140, 141,
144, 208, 210, *217*, *218*
Price, I., 67, *81*
Pridham, J. B., 3, *5*, 84, 85, 86, 88, 90,
92, 94, *95*, *96*
Pubols, M. H., 56, *59*
Pusztai, A., 248, 249, *251*, *252*

Q

Quillet, M., 48, *59*

R

Rabinowitch, E. I., 20, *26*
Racker, E., 36, *46*
Ramachandramurthy, P., 241, 243, 244,
252
Ramus, J., 212, *217*
Ranson, S. L., 22, 23, *25*, *26*
Raschke, W. C., 226, *232*
Ray, Peter M., 162, *163*
Rayle, D. L., 162, *164*
Rebenich, P., 130, 131, 141, *144*
Rees, D. A., 2, 3, *5*, 151, 159, *163*, *164*,
213, *217*
ap Ress, T., 27, 29, 30, 31, 32, 33, 34,
35, 36, *45*, *46*
Reid, J. S. G., 2, 4, *5*, *6*
Reinhold, L., 162, *163*
Reisfeld, R. A., 241, *250*, *252*
Resisener, H. J., 220, 222, *232*
Reyes, E., 221, *232*
Reynolds, D. M., 220, *232*
Ricardo, C. P. P., 34, *46*
Risley, E. B., 176, *180*
Robbins, P. W., 199, 201, *204*, 210, *217*
Roberts, K., 168, 174, 177, 178, *181*
Roberts, R. M., 169, *180*, *181*, 194, *204*,
250, *252*
Robyt, J. F., 109, *125*
Roelofsen, P. A., 159, *164*
Roerig, S., 155, *164*

Subject Index

A

2-Acetamido-2-deoxy-D-galactose, 219, 220, 236, 243, 245

2-Acetamido-2-deoxy-D-glucose, 219–221, 226, 231, 236, 239–240, 242–243, 245

Acetyl CoA, 28–31

N-Acetylgalactosamine, see 2-acetamide-2-deoxy-D-galactose

N-Acetylglucosamine, see 2-acetamido-2-deoxy-D-glucose

N-acetyl-D-glucosaminyl-β-L-asparagine, 239–240, 243

Adenosine 3′-phosphate-5′-phosphosulphate, 212

S-Adenosyl L-methionine, 193–200, 212–213

ADP-glucose, 75–79, 117, 128, 140–141, 208–210

dADP-glucose, 210

ADP-glucose: starch transglucosylase, see starch synthetase

ADP-glucose pyrophosphorylase, 75–89, 140, 208

Agarose, 3

Agglutinin, 237, 238, 240–244

4-L-Alanyl-D-xylopyranose, 240

Alcohol dehydrogenase, 36, 38

Alditols, 47–59

Aldobiouronic acid, 199

Aldose reductase, 56

Aleurone layer, 4, 115, 121

Alginate, 49, 209, 211, 213–216

Allergens, 237, 246–247

Allitol, 48, 50–52, 58

Allulose, 51, 58

Amino sugars, 219–231, 236

2-Amino-2-deoxy-D-glucose, 221, 222, 236, 238, 243–245, 247–248

AMP, 45, 118

Amylases, 109–124, 128, 134, 136, 141–142

maltose, inhibition by, 136–139, 141–143

α-Amylase, 73, 110–124, 230, 231, 245

β-Amylase, 110–124

β-Amylolysis limit, 112, 119, 122

Amylopectin, 111–113, 119, 121–123, 131–132

limit dextrin from, 112, 120

Amylopectin 6-glucanohydrolase 119, 121

Amylose, 111, 113–114, 123, 131–132

3,6-Anhydrogalactose, 212–213

O-Antigens, 151–152, 199, 200–201

Arabinitol, 47, 53–56

Arabinitol dehydrogenase, 56

Arabinogalactan, 150–151, 154–158

Arabinose, 147, 153, 154, 169–172, 175, 177, 197, 198, 219, 239, 244, 248

Arabinose transferase, 198

Ascorbic acid oxidase, 245

Asparagine, 243

Aspartate, 21, 23

Aspartic amino transferase, 66

N-(β-aspartyl)-β-D-(N-acetyl)glucosaminide link, 223

ATP, 8, 11, 16

gluconeogenesis, regulation by, 31, 44–45

ATPase, 163

Auxin, cell elongation and, 160–163

hydrogen ion pump, regulation by, 162–163

B

Bacterial cell wall, 201

Benson-Calvin cycle, see Photosynthesis

Bicarbonate-CO_2 pump, chloroplast, in, 15–16

Branching enzyme, 117

Bromelain, 237, 238, 244–245

Bukuryo, 228